Name _____ Class_____

Skills Worksheet

Concept Review

MW00837568

Section: How are Elements Organized?

Answer the following questions in the space provided.

1. Why do Li, Na, K, Rb, Cs, and Fr all react with Cl in a 1:1 ratio forming substances with similar properties?

_____

_____

_____

2. Explain the method that John Newlands used to organize the elements.

_____

_____

_____

3. What method did Dmitri Mendeleev use to arrange his periodic table?

_____

_____

_____

4. Why did Mendeleev have gaps in his table? How did he use these gaps?

_____

_____

_____

5. What was Henry Moseley's contribution to the periodic table?

_____

_____

_____

| Concept Review *continued*

6. Why was Moseley able to resolve the discrepancies in Mendeleev's table when Mendeleev could not?

_____

_____

_____

7. Explain the importance of valence electrons.

_____

_____

_____

8. Why do elements with similar properties appear at regular intervals in the periodic table?

_____

_____

_____

9. How is the electron configuration similar for each element in a group?

_____

_____

_____

10. How is the electron configuration similar for each element in a period?

_____

_____

_____

Name _____ Class _____ Date _____

Concept Review

Section: Tour of the Periodic Table

Complete each statement below by choosing a term from the following list. Terms may be used more than once.

main-group elements	halogens	metals	transition metals
alkaline earth metals	alkali metals	hydrogen	noble gases

1. The _____ have a single electron in the highest occupied energy level.

2. The _____ are in the s- and p-blocks of the periodic table.

3. All the _____ have two valence electrons and get to a stable electron configuration by losing two electrons.

4. Unlike the main-group elements, each group of the _____ does not have the identical outer electron configuration.

5. The _____, the most reactive group of non-metals, achieve stable electron configurations by gaining one electron.

6. The _____ have a full set of electrons in their outermost energy level.

7. The _____ are very stable and have low reactivity.

8. The _____ are highly reactive and readily form salts with metals.

9. In general, the _____ are metals that are less reactive than the alkali metals and the alkaline earth metals.

10. The _____ are metals that lose one electron when they react with water to form alkaline solutions.

11. Most elements are _____.

12. With its one valence electron, _____ reacts with many other elements.

Answer the following questions in the space provided.

13. Which groups compose the main-group elements?

_____

_____

14. Why are the main-group elements called the *representative elements?*

_____

_____

15. Why are Group 2 elements less reactive than Group 1 elements?

_____

_____

16. Explain why a helium atmosphere is used in welding instead of an oxygen-rich atmosphere.

_____

_____

17. Using electron configurations, explain why the halogens readily react with the alkali metals to form salts.

_____

_____

18. Why is an iron alloy, such as steel, preferred over pure iron?

_____

_____

Complete each statement below by writing the correct word or words in the space provided.

19. The _____ include all members of Groups 1 through 12,

as well as some of the elements of Groups _____ through

_____.

20. Elements in Groups _____ through

_____, including the two long rows below the table, are

called transition elements.

21. In the transition elements, electrons are usually added to the

_____ orbital, which is why these elements are also

known as the _____.

| Concept Review *continued*

22. The _____ include all of the elements in Groups 17 and 18

as well as some members of Groups _____ through

_____.

23. In the _____, electrons are being added to the 4*f* orbitals.

24. In the _____, electrons are being added to the 5*f* orbitals.

25. The _____ are unique in that all are unstable and

radioactive.

Concept Review

Section: Trends in the Periodic Table

Complete each statement below by writing the correct word or words in the space provided.

1. The amount of energy needed to remove an electron from a specific atom is called the _____ energy of the atom.

2. The _____ is half the distance from center to center of two like atoms bonded together.

3. _____ is the energy change that occurs when a neutral atom gains an electron.

4. _____ is a numerical value that reflects how much an atom in a molecule attracts electrons.

5. As the nuclear charge increases across a period, the effective nuclear charge _____ pulling the electrons closer to the nucleus and _____ the size of the atom.

Circle the letter of the choice that best answers the question.

6. Which of the following elements has the largest atomic radius?
   a. boron
   b. aluminum
   c. gallium
   d. indium

7. Which of the following elements has the smallest ionization energy?
   a. potassium
   b. arsenic
   c. nitrogen
   d. bismuth

8. Which of the following elements has the largest electronegativity?

   a. lithium

   b. carbon

   c. chlorine

   d. iodine

Answer the following questions in the space provided.

9. Explain why the exact size of an atom is difficult to determine.

_____

_____

_____

10. Which metal has the larger radius, Li or Na? Why?

_____

_____

11. What is electron shielding?

_____

_____

_____

12. Explain the large decrease in atomic radii as you move across a period from Group 1 to Group 14.

_____

_____

_____

_____

13. Explain why ionization energies tend to decrease down a group.

_____

_____

_____

_____

| Concept Review *continued*

14. Explain the large increase in electronegativity as you move across a period.

_____

_____

_____

_____

15. Complete the following table.

	General Trends	
	From left to right across a period	**Down a group**
Ionization energy		
Atomic radius		
Electronegativity		
Ionic size		
Electron affinity		

Skills Worksheet

Concept Review

Section: Where Did the Elements Come From?

Complete each statement below by writing the correct terms or terms.

1. Most of the atoms in living things come from just six elements,

_____, _____,

_____, _____,

_____, and _____.

2. Immediately after the big bang, temperatures were extremely high and only

_____ could exist.

3. As the universe began to cool, energy was converted to

_____, in the form of _____,

_____, and _____.

4. As the universe continued to cool, these particles joined and formed the first

two elements, _____ and _____.

5. The temperatures in stars get high enough to fuse _____

nuclei with one another, forming elements of still higher atomic numbers.

6. Massive atoms such as iron and nickel form by repeated

_____.

7. When a massive star has converted almost all its core hydrogen and helium

into heavier elements, it collapses and blows apart in an explosion called a

_____ forming elements heavier than iron.

8. The nuclear reaction that changes one nucleus into another by radioactive

disintegration or by bombardment with other particles is called

_____.

| Concept Review *continued*

9. Elements that chemists have created are called _____

   elements.

10. The special equipment that scientists use to create elements are called

    _____.

Answer the following questions in the space provided.

11. There are 93 naturally occurring elements, yet the periodic table contains 113
    elements. Briefly explain the difference in the two numbers.

    _____

    _____

    _____

12. Why are there limits to the synthetic elements that a cyclotron can produce?

    _____

    _____

    _____

13. How does a synchrotron accelerate particles to create synthetic elements?

    _____

    _____

    _____

14. What is the difficulty in identifying superheavy elements?

    _____

    _____

    _____

Assessment

Quiz

Section: How Are Elements Organized?

In the space provided, write the letter of the term or phrase that best answers the question.

_____ **1.** In developing his periodic table, Mendeleev listed on cards each element's name, atomic mass, and
   a. atomic number.
   b. electron configuration.
   c. isotopes.
   d. properties.

_____ **2.** Mendeleev's periodic table did not always list elements in order of increasing atomic mass because he grouped together elements with similar
   a. properties.
   b. atomic numbers.
   c. densities.
   d. colors.

_____ **3.** Mendeleev predicted that the gaps in his periodic table represented
   a. isotopes.
   b. radioactive elements.
   c. permanent gaps.
   d. undiscovered elements.

_____ **4.** The person whose work led to a periodic table based on increasing atomic number was
   a. Moseley.
   b. Mendeleev.
   c. Rutherford.
   d. Cannizzaro.

_____ **5.** An electron that is found in the outermost shell of an atom and determines the atom's chemical properties is called a(n)
   a. valence electron.
   b. paired electron.
   c. p electron.
   d. octave electron.

_____ **6.** The periodic law states that the physical and chemical properties of elements are periodic functions of their atomic
   a. masses.
   b. numbers.
   c. radii.
   d. structures.

_____ **7.** Refer to a periodic table. In which period is calcium?
   a. Period 2
   b. Period 4
   c. Period 6
   d. Period 8

_____ **8.** Refer to a periodic table. In which group is calcium?
   a. Group 1
   b. Group 2
   c. Group 17
   d. Group 18

_____ **9.** An element that has the electron configuration $[Ne]3s^23p^5$ is in which period?
   a. Period 2
   b. Period 3
   c. Period 5
   d. Period 7

_____ **10.** An element that has the electron configuration $[Ne]3s^23p^5$ is in which group?
   a. Group 2
   b. Group 5
   c. Group 7
   d. Group 17

Name _____ Class _____ Date _____

Quiz

Section: Tour of the Periodic Table

In the space provided, write the letter of the term or phrase that best answers the question.

_____ **1.** Elements in the *s*- or *p*-blocks of the periodic table are called
   a. alloys.
   b. main-group elements.
   c. metals.
   d. transition metals.

_____ **2.** Elements in Group 18 have
   a. very low reactivity.
   b. good conductivity.
   c. very high reactivity.
   d. metallic character.

_____ **3.** Nonmetallic elements in Group 17 that react with metals to form salts are
   a. alkali-metals.
   b. halogens.
   c. lanthanides.
   d. noble gases.

_____ **4.** The outer shell electron configuration of an alkaline-earth metal has
   a. one electron in the *s* orbital.
   b. two electrons in the *s* orbital.
   c. one electron in the *p* orbital.
   d. two electrons in the *p* orbital.

_____ **5.** The alkali metals are found on Earth only in compounds because they
   a. have small atoms.
   b. are very reactive elements.
   c. are rare elements.
   d. are metallic elements.

_____ **6.** To which group does hydrogen belong?
   a. Group 1
   b. Group 2
   c. Group 18
   d. None of the above

Quiz *continued*

_____ **7.** A metal is expected to be a(n)
   a. nonconductor.
   b. insulator.
   c. conductor.
   d. fluid at room temperature.

_____ **8.** An element found in Groups 3–12 of the periodic table is classified as a(n)
   a. alkali metal.
   b. alloy.
   c. transition metal.
   d. actinide.

_____ **9.** An element that has an outer shell electron configuration consisting of two electrons in the *d* orbital and one electron in the *s* orbital is in which group?
   a. Group 1
   b. Group 2
   c. Group 3
   d. Group 5

_____ **10.** Lanthanide elements are found in the _____–block of the periodic table.
   a. *s*
   b. *p*
   c. *d*
   d. *f*

Assessment

Quiz

Section: Trends in the Periodic Table

In the space provided, write the letter of the term or phrase that best answers the question.

_____ **1.** Ionization energy is the energy required to remove ____ from an atom of an element.
   a. the electron cloud
   b. all electrons
   c. one electron
   d. an ion

_____ **2.** Across a period in the periodic table, ionization energy generally
   a. decreases.
   b. decreases and then increases.
   c. increases.
   d. remains constant.

_____ **3.** The change in ionization energy down a group is due to
   a. increased electron shielding.
   b. decreased charge of the nucleus.
   c. increased neutrons in the nucleus.
   d. Both (a) and (b)

_____ **4.** When determining the size of an atom by measuring the bond radius, the radius of an atom is
   a. equal to the distance between nuclei.
   b. one-half the distance between nuclei.
   c. twice the distance between nuclei.
   d. one-fourth the distance between nuclei.

_____ **5.** Across a period in the periodic table, atomic radii generally
   a. decrease.
   b. decrease, then increase.
   c. increase.
   d. increase, then decrease.

_____ **6.** Down a group in the periodic table, atomic radii generally
   a. decrease.
   b. remain constant.
   c. increase.
   d. vary unpredictably.

Quiz *continued*

_____ **7.** An element with the lowest electronegativity would be found in ____ of the periodic table.
  a. Group 1, Period 7
  b. Group 3, Period 4
  c. Group 5, Period 3
  d. Group 17, Period 2

_____ **8.** Refer to a periodic table and determine which element has the lowest electron affinity.
  a. Cl
  b. Se
  c. Cs
  d. Te

_____ **9.** As the atomic number of the metals of Group 1 increases, the ionic radius
  a. increases.
  b. decreases.
  c. remains the same.
  d. cannot be determined.

_____ **10.** An element with the smallest anionic (negative-ionic) radius would be found in ____ of the periodic table.
  a. Group 1, Period 7
  b. Group 3, Period 4
  c. Group 5, Period 3
  d. Group 17, Period 2

Assessment

Quiz

Section: Where Did the Elements Come From?

In the space provided, write the letter of the term or phrase that best answers the question.

_____ **1.** The currently accepted model of the universe's beginnings includes which of the following hypotheses?
  a. Hydrogen and helium were the first elements formed.
  b. The universe started with an event called the big bang.
  c. As the universe expanded, it cooled and some of its initial energy was converted into matter.
  d. All of the above

_____ **2.** The first stars formed in clouds of
  a. helium.
  b. oxygen.
  c. hydrogen.
  d. carbon.

_____ **3.** The principal source of a star's energy is
  a. nuclear fusion.
  b. gravitational attraction within a star.
  c. uranium.
  d. the synthesis of heavy metals.

_____ **4.** Within stars, energy is released when two helium nuclei fuse forming an element that has an atomic number of
  a. 2.
  b. 4.
  c. 8.
  d. 16.

_____ **5.** The production of an oxygen ion and a proton by the collision of an alpha particle and a nitrogen atom is an example of a(n)
  a. nuclear reaction.
  b. transmutation.
  c. super collision.
  d. Both a and b

_____ **6.** Which element is synthetic?
  a. uranium
  b. iodine
  c. helium
  d. nobelium

_____ **7.** Ninety-three elements occur in nature; the other known elements have been
　　a. found in interstellar space.
　　b. found in the atmosphere.
　　c. synthesized.
　　d. found in Earth's core.

_____ **8.** Because a particle accelerator supplies energy to charged particles, it would NOT be able to accelerate
　　a. electrons.
　　b. protons.
　　c. neutrons.
　　d. Both a and b

_____ **9.** A synchrotron achieves greater particle speeds than a cyclotron because the synchrotron
　　a. can supply more energy.
　　b. supplies pulses of energy to the particles.
　　c. compensates for particle mass increase.
　　d. All of the above

_____ **10.** Superheavy elements
　　a. are found in Earth's core.
　　b. are members of period 3 in the periodic table.
　　c. exist for centuries.
　　d. have atomic numbers greater than 106.

Name _____ Class _____ Date _____

Chapter Test

The Periodic Table

Use the periodic table above to answer the questions in this Chapter Test.

In the space provided, write the letter of the term or phrase that best completes each statement or best answers each question.

_____ **1.** Mendeleev attempted to organize the chemical elements based on their
   a. symbols.
   b. properties.
   c. atomic numbers.
   d. electron configurations.

_____ **2.** A horizontal row in the periodic table is called a(n)
   a. family.
   b. group.
   c. octet.
   d. period.

_____ **3.** The periodic law states that
 a. no two electrons with the same spin can be found in the same place in an atom.
 b. the physical and chemical properties of the elements are functions of their atomic number.
 c. electrons exhibit properties of both waves and particles.
 d. the chemical properties of elements can be grouped according to periodicity.

_____ **4.** An element with the general electron configuration for its outermost electrons of ns^2np^1 would be in which element group?
 a. Group 2
 b. Group 13
 c. Group 14
 d. Group 15

_____ **5.** How many electrons does carbon have in its outermost shell?
 a. 1
 b. 2
 c. 3
 d. 4

_____ **6.** Which of the following elements behaves similarly to calcium?
 a. magnesium
 b. sodium
 c. sulfur
 d. chlorine

_____ **7.** Which periodic group or series of elements is not correctly matched with its common family name?
 a. alkaline-earth metals   Group 2
 b. transition metals   Group 3
 c. halogens   Group 17
 d. noble gases   Group 18

_____ **8.** Highly reactive metallic elements that react with water to form alkaline solutions are called
 a. actinides.
 b. alkali metals.
 c. halogens.
 d. noble gases.

_____ **9.** The electron configurations of main-group elements end in
    a. d and f orbitals.
    b. s and p orbitals.
    c. s and d orbitals.
    d. p and d orbitals.

_____**10.** Which of the following elements is a transition metal?
    a. calcium
    b. iron
    c. sodium
    d. sulfur

_____**11.** Which property is not characteristic of a metal?
    a. conductivity
    b. brittleness
    c. luster
    d. ductility

_____**12.** The alkali metal elements are found in _____ of the periodic table.
    a. Group 1
    b. Group 2
    c. Period 1
    d. Period 2

_____**13.** The energy required to remove an electron from an atom is called the atom's
    a. electron affinity.
    b. electron energy.
    c. electronegativity.
    d. ionization energy.

_____**14.** A measure of the ability of an atom in a chemical compound to attract electrons is called
    a. electron affinity.
    b. electron configuration.
    c. electronegativity.
    d. ionization potential.

_____**15.** Which of the following elements has the greatest atomic radius?
    a. Al
    b. S
    c. Si
    d. C

_____**16.** Which of the following elements has the lowest electronegativity?
  a. C
  b. F
  c. Li
  d. O

_____**17.** Which of the following elements has the greatest ionization energy?
  a. Ga
  b. K
  c. Bi
  d. As

_____**18.** Which of the following elements has the greatest electron affinity (largest positive value)?
  a. Br
  b. As
  c. Ar
  d. I

_____**19.** Naturally occurring elements are created in stars by a process known as
  a. electrolysis.
  b. fission.
  c. transmutation.
  d. fusion.

_____**20.** Many elements have been synthesized by chemists in laboratories. These elements are the
  a. noble gases.
  b. alkaline-earth metals.
  c. transuranium metals.
  d. lanthanides.

| Chapter Test *continued*
_____

Answer the questions in the spaces provided.

21. Why are the noble gases not reactive?

_____

_____

_____

_____

22. How does a cyclotron work?

_____

_____

_____

Answer the question on a separate piece of paper.

23. Explain what happens when a star uses all of its hydrogen and helium.

Answer each of the following problems in the spaces provided.

24. Arrange the following elements in order of decreasing ionization energy.

Rb, In, Sn, Sb, As

25. Arrange the following elements in order of increasing atomic radius.

Sr, Rb, Sb, I, In

26. Over on paper: How does atomic radius change going down a Group. Explain why this happens

Skills Practice Lab

The Mendeleev Lab of 1869

Russian chemist Dmitri Mendeleev is generally credited as being the first chemist to observe that patterns emerge when the elements are arranged according to their properties. Mendeleev's arrangement of the elements was unique because he left blank spaces for elements that he claimed were undiscovered as of 1869. Mendeleev was so confident that he even predicted the properties of these undiscovered elements. His predictions were eventually proven to be quite accurate, and these new elements fill the spaces that originally were blank in his table.

Use your knowledge of the periodic table to determine the identity of each of the nine unknown elements in this activity. The unknown elements are from the groups in the periodic table that are listed below. Each group listed below contains at least one unknown element.

1   2   11   13   14   17   18

None of the known elements serves as one of the nine unknown elements. No radioactive elements are used during this experiment. The relevant radioactive elements include Fr, Ra, At, and Rn. You may not use your textbook or other reference materials. You have been provided with enough information to determine each of the unknown elements.

OBJECTIVES

Observe the physical properties of common elements.

Observe the properties and trends in the elements on the periodic table.

Draw conclusions and identify unknown elements based on observed trends in properties.

MATERIALS

- blank periodic table
- elemental samples: Ar, C, Sn, Pb
- note cards, 3 in. $\times$ 5 in.
- periodic table

Always wear safety goggles and a lab apron to protect your eyes and clothing. If you get a chemical in your eyes, immediately flush the chemical out at the eyewash station while calling to your teacher. Know the location of the emergency lab shower and eyewash station and the procedures for using them.

| The Mendeleev Lab of 1869 *continued*

Procedure

1. Put on safety goggles and a lab apron.

2. Use the note cards to copy the information listed on each of the sample cards. If the word *observe* is listed, you will need to visually inspect the sample and then write the observation in the appropriate space.

3. Arrange the note cards of the known elements in a crude representation of the periodic table. In other words, all of the known elements from Group 1 should be arranged in the appropriate order. Arrange all of the other cards accordingly.

4. Once the cards of the known elements are in place, inspect the properties of the unknowns to see where their properties would best "fit" the trends of the elements of each group.

5. Assign the proper element name to each of the unknowns. Add the symbol for each one of the unknown elements to your data table.

6. Clean up your lab station, and return the leftover note cards and samples of the elements to your teacher. Do not pour any of the samples down the drain or in the trash unless your teacher directs you to do so. Wash your hands thoroughly before you leave the lab and after all your work is finished.

TABLE 1 UNKNOWN ELEMENTS IDENTIFICATION

Unknown #	Element	Unknown #	Element
1		6	
2		7	
3		8	
4		9	
5			

Conclusions

1. **Interpreting information** Summarize your group's reasoning for the assignment of each unknown. Explain in a few sentences exactly how you predicted the identity of the nine unknown elements.

_____

_____

The Mendeleev Lab of 1869 *continued*

Li	
Physical state	solid
Density	0.534 g/cm^3
Hardness	soft, claylike
Conductivity	good
Melting point	180°C
Solubility (H$_2$O)	reacts with water
Color	silver

Ag	
Physical state	solid
Density	10.50 g/cm^3
Hardness	somewhat soft
Conductivity	excellent
Melting point	961°C
Solubility (H$_2$O)	none
Color	silver

Cu	
Physical state	(observe)
Density	8.96 g/cm^3
Hardness	somewhat soft
Conductivity	excellent
Melting point	1083°C
Solubility (H$_2$O)	none
Color	(observe)

C	
Physical state	(observe)
Density	2.10 g/cm^3
Hardness	soft yet brittle
Conductivity	good
Melting point	3550°C
Solubility (H$_2$O)	negligible
Colo	(observe)

Cl$_2$	
Physical state	gas
Density	0.00321 g/cm^3
Hardness	none
Conductivity	very poor
Melting point	−101°C
Solubility (H$_2$O)	slight
Color	greenish yellow

He	
Physical state	gas
Density	0.00018 g/cm^3
Hardness	none
Conductivity	very poor
Melting point	−272°C
Solubility (H$_2$O)	none
Color	colorless

Na	
Physical state	solid
Density	0.971 g/cm^3
Hardness	soft, claylike
Conductivity	good
Melting point	98°C
Solubility (H$_2$O)	reacts rapidly
Color	silver

Ca	
Physical state	solid
Density	1.57 g/cm^3
Hardness	medium
Conductivity	good
Melting point	845°C
Solubility (H$_2$O)	reacts
Color	silvery white

Be	
Physical state	solid
Density	1.85 g/cm^3
Hardness	brittle
Conductivity	excellent
Melting point	1287°C
Solubility (H$_2$O)	none
Color	gray

Sn	
Physical state	(observe)
Density	7.31 g/cm^3
Hardness	somewhat soft
Conductivity	good
Melting point	232°C
Solubility (H$_2$O)	none
Color	(observe)

Ne	
Physical state	gas
Density	0.00090 g/cm^3
Hardness	none
Conductivity	very poor
Melting point	−249°C
Solubility (H$_2$O)	none
Color	colorless

Br$_2$	
Physical state	liquid
Density	3.12 g/cm^3
Hardness	none
Conductivity	very poor
Melting point	−7.2°C
Solubility (H$_2$O)	negligible
Color	reddish brown

K	
Physical state	solid
Density	0.86 g/cm^3
Hardness	soft, claylike
Conductivity	good
Melting point	63°C
Solubility (H$_2$O)	reacts rapidly
Color	silver

Ba	
Physical state	solid
Density	3.6 g/cm^3
Hardness	soft
Conductivity	good
Melting point	710°C
Solubility (H$_2$O)	reacts strongly
Color	silvery white

Xe	
Physical state	gas
Density	0.00585 g/cm^3
Hardness	none
Conductivity	very poor
Melting point	−111.9°C
Solubility (H$_2$O)	none
Color	colorless

Name _____ Class _____ Date _____

In	
Physical state	solid
Density	7.31 g/cc
Hardness	very soft
Conductivity	medium
Melting point	157°C
Solubility (H_2O)	none
Color	silvery white

I_2	
Physical state	solid
Density	4.93 g/cm^3
Hardness	soft
Conductivity	very poor
Melting point	113.5°C
Solubility (H_2O)	negligible
Color	bluish-black

Pb	
Physical state	(observe)
Density	11.35 g/cm^3
Hardness	somewhat soft
Conductivity	poor
Melting point	327.5°C
Solubility (H_2O)	none
Color	(observe)

Ar	
Physical state	(observe)
Density	0.00178 g/cm^3
Hardness	none
Conductivity	very poor
Melting point	−189.2°C
Solubility (H_2O)	none
Color	(observe)

Ga	
Physical state	solid
Density	5.904 g/cc
Hardness	soft
Conductivity	medium
Melting point	30°C
Solubility (H_2O)	none
Color	silvery

Cs	
Physical state	solid
Density	1.87 g/cm^3
Hardness	soft
Conductivity	good
Melting point	29°C
Solubility (H_2O)	reacts violently
Color	silvery white

Unknown #1	
Physical state	solid
Density	2.33 g/cm^3
Hardness	brittle
Conductivity	intermediate
Melting point	1410°C
Solubility (H_2O)	none
Color	gray

Unknown #2	
Physical state	gas
Density	0.00170 g/cm^3
Hardness	none
Conductivity	very poor
Melting point	−219.6°C
Solubility (H_2O)	slight
Color	pale yellow

Unknown #3	
Physical state	solid
Density	1.53 g/cm^3
Hardness	soft
Conductivity	good
Melting point	39°C
Solubility (H_2O)	reacts violently
Color	silvery white

Unknown #4	
Physical state	gas
Density	0.00374 g/cm^3
Hardness	none
Conductivity	very poor
Melting point	−156.6°C
Solubility (H_2O)	none
Color	colorless

Unknown #5	
Physical state	solid
Density	19.3 g/cm^3
Hardness	soft
Conductivity	excellent
Melting point	1064°C
Solubility (H_2O)	none
Color	gold

Unknown #6	
Physical state	solid
Density	2.54 g/cm^3
Hardness	somewhat soft
Conductivity	good
Melting point	769°C
Solubility (H_2O)	reacts rapidly
Color	silvery white

Unknown #7	
Physical state	solid
Density	5.32 g/cm^3
Hardness	fairly brittle
Conductivity	fair to poor
Melting point	937°C
Solubility (H_2O)	none
Color	gray

Unknown #8	
Physical state	solid
Density	1.74 g/cm^3
Hardness	medium
Conductivity	good
Melting point	651°C
Solubility (H_2O)	reacts slowly
Color	silvery white

Unknown #9	
Physical state	solid
Density	11.85 g/cc
Hardness	very soft
Conductivity	medium
Melting point	303°C
Solubility (H_2O)	none
Color	silvery white

Reactivity of Halide Ions

The four halide salts used in this experiment are found in your body. Although sodium fluoride is poisonous, trace amounts seem to be beneficial to humans in the prevention of tooth decay. Sodium chloride is added to most of our food to increase flavor while masking sourness and bitterness. Sodium chloride is essential for many life processes, but excessive intake appears to be linked to high blood pressure. Sodium bromide is distributed throughout body tissues, and in the past it has been used as a sedative. Sodium iodide is necessary for the proper operation of the thyroid gland, which controls cell growth. The concentration of sodium iodide is almost 20 times greater in the thyroid than in blood. The need for this halide salt is the reason that about 10 ppm of NaI is added to packages of table salt labeled "iodized."

The principal oxidation number of the halogens is −1. However, all halogens except fluorine may have other oxidation numbers. The specific tests you will develop in this experiment involve the production of recognizable precipitates and complex ions. You will use your observations to determine the halide ion present in an unknown solution.

MATERIALS

- 24-well microplate
- $AgNO_3$, 0.1 M
- $Ca(NO_3)_2$, 0.5 M
- gloves
- KBr, 0.2 M
- KI, 0.2 M
- lab apron
- $Na_2S_2O_3$, 0.2 M

- NaCl, 0.1 M
- NaF, 0.1 M
- NaOCl (commercial bleach), 5%
- $NH_3(aq)$, 4 M
- safety goggles
- starch solution, 3%
- thin-stemmed pipets (12)

Always wear safety goggles and a lab apron to protect your eyes and clothing. If you get a chemical in your eyes, immediately flush the chemical out at the eyewash station while calling to your teacher. Know the location of the emergency lab shower and eyewash station and the procedures for using them.

Do not touch any chemicals. If you get a chemical on your skin or clothing, wash the chemical off at the sink while calling to your teacher. Make sure you carefully read the labels and follow the precautions on all containers of chemicals that you use. If there are no precautions stated on the label, ask your teacher what precautions to follow. Do not taste any chemicals or items used in the laboratory. Never return leftovers to their original container; take only small amounts to avoid wasting supplies.

Call your teacher in the event of a spill. Spills should be cleaned up promptly, according to your teacher's directions.

| Reactivity of Halide Ions *continued*

Acids and bases are corrosive. If an acid or base spills onto your skin or clothing, wash the area immediately with running water. Call your teacher in the event of an acid spill. Acid or base spills should be cleaned up promptly.

 Never put broken glass in a regular waste container. Broken glass should be disposed of separately according to your teacher's instructions.
   Never stir with a thermometer because the glass around the bulb is fragile and might break.

OBJECTIVES

Observe the reactions of the halide ions with different reagents.

Analyze data to determine characteristic reactions of each halide ion.

Infer the identity of unknown solutions.

Procedure

1. Put on safety goggles, gloves, and a lab apron.

2. Put 5 drops of 0.1 M NaF into each of four wells in row A, as shown in **Figure 1.** Put 5 drops of 0.1 M NaCl into each of the wells in row B. Put 5 drops of 0.2 M KBr into each of the wells in row C and 5 drops of 0.2 M KI into each of the wells in Row D. Reserve rows E and F for unknown solutions.

Figure 1

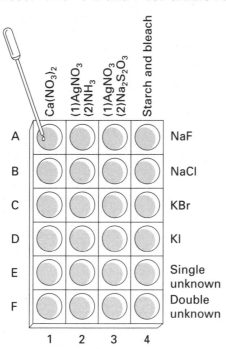

3. Add 5 drops of 0.5 M $Ca(NO_3)_2$ solution to each of the four halide solutions in column 1. Record your observations in the **Table 1.**

4. Add 2 drops of 0.1 M $AgNO_3$ solution to each of the halides in columns 2 and 3. Record in **Table 1** the colors of the precipitates formed.

5. Add 5 drops of 4 M $NH_3(aq)$ to the precipitates in column 2. Record your observations in the **Table 1.**

6. Add 5 drops of 0.2 M $Na_2S_2O_3$ solution to the precipitates in column 3. Record your observations in **Table 1.**

7. To the halides in column 4, add 5 drops of starch solution and 1 drop of 5% bleach solution. Record your observations. Save the results of testing the known halide solutions for comparison with the tests of the unknown solutions.

8. Obtain an unknown solution. Put 5 drops of the unknown in each of the four wells in row E. Add the reagents to each well as you did in steps 3–6. Compare the results with those of the known halides in rows A–D. Record your findings in **Table 1,** and identify the unknown.

9. Obtain an unknown solution containing a mixture of two halide ions. Place 5 drops of the unknown mixture in each of the four wells in row F. Add the reagents to each well as you did in steps 3–6. Record your results. Compare the results with those of the known halides in rows A–D. Identify the halides in the double unknown solution.

10. Rinse the microplate into a trough or dishpan provided by your teacher. Clean all apparatus and your lab station. Return equipment to its proper place. Dispose of chemicals and solutions in the containers designated by your teacher. Do not pour any chemicals down the drain or in the trash unless your teacher directs you to do so. Wash your hands thoroughly before you leave the lab and after all work is finished.

TABLE 1: RESULTS OF THE REACTIONS OF HALIDE SALTS

Halide salts	$Ca(NO_3)_2$	$AgNO_3$	$AgNO_3$ + NH_3	$AgNO_3$ + $Na_2S_2O_3$	$NaOCl$ + starch
NaF					
NaCl					
KBr					
KI					
Single unknown					
Double unknown					

Reactivity of Halide Ions *continued*

Analysis

1. **Analyzing Data** Which procedure(s) confirm(s) the presence of (a) F^- ions, (b) Cl^- ions, (c) Br^- ions, (d) I^- ions?

Conclusions

1. **Drawing Conclusions** What generalizations can be made about silver halides?

   _____

   _____

2. **Applying Conclusions** In nuclear explosions or accidents, iodine-131, a radioactive fission product, can become dispersed in the atmosphere. Eventually, the iodine isotope will fall onto the ground and be absorbed by plants. Explain how radiation from iodine-131 could become concentrated in the human body and cause a growth disorder.

   _____

   _____

   _____

   _____

3. **Defending Conclusions** Identify your unknown(s) and use your experimental evidence to support your identifications.

   _____

   _____

   _____

Exploring the Periodic Table

The periodic table of elements, found in your chemistry textbook, contains a wealth of information. Within the periodic table, each of the known elements is listed by name, as well as symbol. Other information found in the periodic table includes the atomic weight along with the atomic number of each element. Each element in the periodic table is arranged according to its properties. For example, the elements listed in the first column of the table are the alkali metals. These elements all have low densities, low melting points, and good electrical conductivity.

The same wealth of information found in the periodic table can be found within the Periodic application on your TI-83 Plus graphing calculator. In this activity, you will learn how to navigate through the Periodic application and access this information. In addition to the information found within the periodic table, the Periodic application contains additional information about each element such as density, melting point, boiling point, electronegativity, and oxidation states.

OBJECTIVES

- **Explore** the periodic table of elements using the Periodic application.
- **Identify** various elements found in the periodic table.
- **Observe** elemental trends using the graphing feature of the Periodic application.
- **Calculate** the atomic weight of various chemical compounds.

EQUIPMENT

- TI-83 Plus graphing calculator
- Periodic application

PRELAB

Navigational menus for the Periodic application can be found at the bottom of the screen.

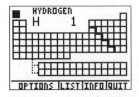

To select one of the menu items found at the bottom of the screen, press the calculator key located directly below that item. For example, to select LIST press ⌈ZOOM⌉, and to select INFO press ⌈TRACE⌉. To select OPTIONS, press either ⌈ Y= ⌉ or ⌈WINDOW⌉ because it has two calculator keys located below it.

Name _____ Class _____ Date _____

To select items from a vertical screen, such as the one shown below, move the highlight cursor up or down through the list by pressing ▲ or ▼. When the option you wish to select is highlighted, press [ENTER] or select OK.

Procedure

1. Turn on the calculator. Press [APPS], and then press the calculator key for the *number* that precedes the PERIODIC application. You are now at the main screen of the program.

TABLE SCREEN

When the Periodic application is started, the table screen is displayed. Follow the instructions below to move through the table and select the element sodium.

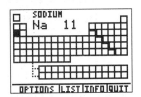

2. Move the cursor from cell to cell by pressing ▲, ▼, ◀, or ▶. Position the cursor over the cell for sodium. When the cursor is positioned properly, the name of the element is displayed at the top of the screen.

3. Press [ENTER] to display detailed properties for the element sodium.

4. Press ▲ or ▼ to scroll through the properties for sodium. To scroll the screen a page at a time, press [ALPHA] and then ▲ or ▼.

When the properties screen is displayed, values for each of the properties can be copied to the calculator Home screen for use in calculations. Follow the instructions below to calculate the atomic mass of NaCl.

5. Press ▼ to move the cursor down until WEIGHT is highlighted.

6. Select SET to copy the value for atomic mass to the calculator Home screen.

7. Select QUIT to exit the Periodic application and press [+].

8. Restart the Periodic application.

9. Repeat Steps 2–4 to select the element chlorine.

10. Repeat Steps 5–6 to copy the atomic mass of chlorine to the calculator Home screen.

11. Select QUIT to exit the Periodic application. Press ⎡ENTER⎤ to calculate the atomic mass of NaCl.

DISPLAYING ELEMENT GROUPS

Groups of elements can be highlighted and displayed on the table screen. Follow the instructions below to highlight the alkali metals on the table screen.

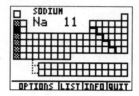

1. From the table screen, select OPTIONS to display the Options menu.

2. Press ⎡ENTER⎤ to select HIGHLIGHT REGIONS from the Options menu.

3. Press ⎡▼⎤ to move the cursor down until ALKALI METALS is highlighted, and press ⎡ENTER⎤. The table screen will now be displayed with the selected area highlighted.

GRAPHING ELEMENTS BY PROPERTIES

To help identify trends within the periodic table, the Periodic application has five graphing options. Each graph has the atomic number displayed on the *x*-axis and the atomic radius, 1st ionization energy, electronegativity, density, or melting point displayed on the *y*-axis. Follow the instructions below to display a graph of atomic number versus density.

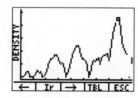

1. From the table screen, select OPTIONS to display the Options menu.

2. Press ⎡▼⎤ to move the cursor down until GRAPH PROPERTIES is highlighted, and press ⎡ENTER⎤.

3. Press ⎡▼⎤ to move the cursor down until DENSITY is highlighted, and press ⎡ENTER⎤. A graph of density versus atomic number will be displayed.

4. Press ⎡◀⎤ or ⎡▶⎤ to move the trace cursor from element to element.

Exploring the Periodic Table *continued*

DISPLAYING ELEMENTS AS A LIST

Instead of the table screen, the elements can be displayed in list format. Once in a list, the elements can be sorted by atomic number, name, or symbol. Follow the instructions below to display a list of the elements sorted by symbol.

```
Ac   ACTINIUM     89
Ag   SILVER       47
Al   ALUMINIUM    13
Am   AMERICIUM    95
Ar   ARGON        18
As   ARSENIC      33
At   ASTATINE     85
          SORT TBL QUIT
```

1. From the table screen, select LIST to display a list of the elements.

2. Select SORT from the list screen.

3. Press ▼ to move the cursor down until SYMBOL is highlighted, and press ENTER. The list of elements will now be sorted according to symbol.

4. To select an element, press ▲ or ▼ to move the cursor from element to element. When your element is highlighted, press ENTER to display properties for that element.

SAMPLE PROBLEMS

1. Using the Periodic application, calculate the atomic mass for each of the following.

  a. Bromine, Br

  b. Tungsten, W

  c. Potassium chloride, KCl

  d. Glucose, $C_6H_{12}O_6$

2. Display the graph of electronegativity versus atomic number. Using the graph, determine what happens to the electronegativity as you move from left to right across the periodic table.

3. Display the graph of density versus atomic number. Using the graph, determine which element has the greatest density. Identify the element, and list its density.

4. Determine the density for each of the following.

  a. argon, Ar

  b. aluminum, Al

  c. strontium, Sr

  d. iron, Fe

5. Using the Highlight Region option, identify the elements that make up the alkali earth metals.

Lesson Plan

Section: How Are Elements Organized?

Pacing

Regular Schedule	with lab(s): 2 days	without lab(s): 1 day
Block Schedule	with lab(s): 1 day	without lab(s): ½ day

Objectives

1. Describe the historical development of the periodic table.

2. Describe the organization of the modern periodic table according to the periodic law.

National Science Education Standards Covered

UNIFYING CONCEPTS AND PROCESSES

UCP 1 Systems, order, and organization

UCP 5 Form and function

PHYSICAL SCIENCE—STRUCTURE AND PROPERTIES OF MATTER

PS 2b An element is composed of a single type of atom. When elements are listed in order according to the number of protons (called the atomic number), repeating patterns of physical and chemical properties identify families of elements with similar properties. This "Periodic Table" is a consequence of the repeating pattern of outermost electrons and their permitted energies.

> **KEY**
> **SE** = Student Edition
> **ATE** = Annotated Teacher Edition

Block 1 *(45 minutes)*

FOCUS *5 minutes*

❑ **Bellringer,** ATE (GENERAL). This activity has students list things in the classroom that they think are made from single elements.

MOTIVATE *5 minutes*

❑ **Identifying Preconceptions,** ATE (GENERAL). This activity helps students understand that elements that have similar chemical properties do not have to have similar physical properties.

Lesson Plan *continued*

TEACH *25 minutes*

❑ **Transparency,** Blocks of the Periodic Table (GENERAL). This transparency master illustrates how the shape of the periodic table is determined by how electrons fill orbitals. (Figure 3)

❑ **Demonstration,** ATE (GENERAL). This demonstration allows students to inspect samples from the *s-*, *p-*, and *d-*blocks of the periodic table.

❑ **Using the Figure,** ATE (GENERAL). Review the different kinds of information shown in Figure 4 of the periodic table of elements.

❑ **Teaching Tip,** ATE (GENERAL). Explain the electron configurations of transition metals.

❑ **Datasheets for In-Text Lab: The Mendeleev Lab of 1869,** SE (GENERAL). Students observe the physical properties of common elements and identify unknown elements based on observed trends in properties.

CLOSE *10 minutes*

❑ **Quiz,** ATE (GENERAL). This assignment has students answer questions about the concepts in this lesson.

❑ **Reteaching,** ATE (BASIC). Ask students to explain why the periodic table is such a valuable tool for scientists. Then, review with students the value of the information in the periodic table.

❑ **Assessment Worksheet: Section Quiz** (GENERAL)

HOMEWORK

❑ **Group Activity,** ATE (GENERAL). Have students each draw one number that corresponds to an element in the periodic table. Students should prepare a paragraph about the element whose number he or she drew.

❑ **Section Review,** SE (GENERAL). Assign items 1–14

❑ **Skills Worksheet: Concept Review** (GENERAL)

❑ **Interactive Tutor for ChemFile,** Module 3: Periodic Properties

OTHER RESOURCES

❑ **Homework,** ATE (ADVANCED). This assignment has students work through Mendeleev's process for predicting the presence of unknown elements.

❑ **Using the Figure,** ATE (GENERAL). Allow students to compare Figure 4 of the periodic table of elements in their textbook with a large wall version of the table.

❑ **go.hrw.com**

❑ **www.scilinks.org**

Lesson Plan

Section: Tour of the Periodic Table

Pacing

Regular Schedule	**with lab(s):** 2 days	**without lab(s):** 1 day
Block Schedule	**with lab(s):** 1 day	**without lab(s):** ½ day

Objectives

1. Locate the different families of main-group elements on the periodic table, describe their characteristic properties, and relate their properties to their electron configurations.

2. Locate metals on the periodic table, describe their characteristic properties, and relate their properties to their electron configurations.

National Science Education Standards Covered
UNIFYING CONCEPTS AND PROCESSES

UCP 1 Systems, order, and organization

UCP 5 Form and function

PHYSICAL SCIENCE—STRUCTURE AND PROPERTIES OF MATTER

PS 2b An element is composed of a single type of atom. When elements are listed in order according to the number of protons (called the atomic number), repeating patterns of physical and chemical properties identify families of elements with similar properties. This "Periodic Table" is a consequence of the repeating pattern of outermost electrons and their permitted energies.

> **KEY**
> **SE** = Student Edition
> **ATE** = Annotated Teacher Edition

Block 2 *(45 minutes)*
FOCUS *5 minutes*

❑ **Bellringer,** ATE (GENERAL). This activity has students label each group of elements on a blank periodic table by the configuration of the elements' valence electrons.

MOTIVATE *5 minutes*

❑ **Using the Figure,** SE (GENERAL). Point out the location of the main-group elements and the metals in Figure 5. Using a wall size periodic table, review the location of groups 1, 2, and 13–18 on the periodic table. Have students name elements that belong to each group.

TEACH *30 minutes*

❑ **Demonstration,** ATE (GENERAL). This demonstration allows you to show students the reactivities of magnesium and calcium.

❑ **Demonstration,** ATE (GENERAL). This demonstration allows you to show students the properties of metals.

❑ **Misconception Alert,** ATE (GENERAL). Help students distinguish the difference between the meaning of the term *metal* as applied to elements in the periodic table and this term as applied to everyday metallic objects.

❑ **Group Activity,** ATE (GENERAL). Students use references to research the properties different groups of elements. Mix up the properties and have students match them with the different groups.

❑ **Microscale Lab: Reactivity of Halide Ions,** Chapter Resource File (GENERAL). Students observe reactions of halide ions with different reagents and analyze data to determine characteristics reactions of each halide ion.

CLOSE *5 minutes*

❑ **Quiz,** ATE (GENERAL). This assignment has students answer questions about the concepts in this lesson.

❑ **Assessment Worksheet: Section Quiz** (GENERAL)

HOMEWORK

❑ **Section Review,** SE (GENERAL). Assign items 1–13.

❑ **Skills Worksheet: Concept Review** (GENERAL)

OTHER RESOURCES

❑ **Demonstration,** ATE (GENERAL). This demonstration allows you to show students the orange-yellow light of sodium.

❑ **Homework,** ATE (ADVANCED). This assignment has students research the gems sapphire and ruby.

❑ **go.hrw.com**

❑ **www.scilinks.org**

Lesson Plan

Section: Trends in the Periodic Table

Pacing

Regular Schedule	**with lab(s):** 3 days	**without lab(s):** 2 days
Block Schedule	**with lab(s):** 1½ days	**without lab(s):** 1 day

Objectives

1. Describe the periodic trends in ionization energy, and relate them to the atomic structures of the elements.

2. Describe periodic trends in atomic radius, and relate them to the atomic structures of the elements.

3. Describe periodic trends in electronegativity, and relate them to the atomic structures of the elements.

4. Describe periodic trends in ionic size, electron affinity, and melting and boiling points, and relate them to the atomic structures of the elements.

National Science Education Standards Covered

UNIFYING CONCEPTS AND PROCESSES

UCP 1 Systems, order, and organization

UCP 5 Form and function

PHYSICAL SCIENCE—STRUCTURE AND PROPERTIES OF MATTER

PS 2b An element is composed of a single type of atom. When elements are listed in order according to the number of protons (called the atomic number), repeating patterns of physical and chemical properties identify families of elements with similar properties. This "Periodic Table" is a consequence of the repeating pattern of outermost electrons and their permitted energies.

KEY
SE = Student Edition
ATE = Annotated Teacher Edition

Block 3 *(45 minutes)*

FOCUS *10 minutes*

❑ **Bellringer,** ATE (GENERAL). Have students draw atomic models of lithium, magnesium, and fluorine and predict whether the ions of these elements will be larger or smaller than the atoms.

Lesson Plan *continued*

MOTIVATE *5 minutes*

❏ **Discussion,** (GENERAL) Write the term *trend* on the board or on a transparency. Ask students to suggest a definition for the term and to give examples of everyday trends, such as in fashion, food, art, architecture, color, and so on. End the discussion with a reading of the definition of the term from the student textbook.

TEACH *30 minutes*

❏ **Using the Figure,** ATE (GENERAL). Students speculate about what might happen if the lithium atom in Figure 16 were in the presence of a fluorine atom.

❏ **Transparency,** Periodic Trends of Radii. (GENERAL) This transparency master illustrates that atomic radius generally increases down a group and decreases across a period. (Figure 20)

HOMEWORK

❏ **Section Review,** SE (GENERAL). Assign items 1–6, 12, 15, 16, and 18.

OTHER RESOURCES

❏ **Homework,** ATE (ADVANCED). This assignment asks students to explain the large jump in the fifth ionization energy for carbon.

❏ **Demonstration,** ATE (ADVANCED). This demonstration allows students to indirectly determine the size of an oleic acid molecule.

❏ **go.hrw.com**

❏ **www.scilinks.org**

Block 4 *(45 minutes)*

TEACH *35 minutes*

❏ **Group Activity,** ATE (GENERAL). This activity allows students to play a card game based on electronegativity numbers.

❏ **Demonstration**, ATE (BASIC). This demonstration helps students visualize ionic size.

❏ **Transparency,** Additional Periodic Trends. This transparency master shows additional periodic trends.

❏ **CBL™ Probeware Lab: Exploring the Periodic Table,** Chapter Resource File (Advanced). Students will learn how to navigate through the Periodic application of the TI-83 Plus and access information about each element, including density, melting point, boiling point, electronegativity, and oxidation states.

Lesson Plan *continued*

CLOSE *10 minutes*

❏ **Reteaching,** ATE (BASIC). Students work in pairs to review periodic trends.

❏ **Quiz,** ATE (GENERAL). This assignment has students answer questions about the concepts in this lesson.

❏ **Assessment Worksheet: Section Quiz** (GENERAL)

HOMEWORK

❏ **Section Review,** SE (GENERAL). Assign items 7–11, 13, 14, and 17.

❏ **Skills Worksheet: Concept Review** (GENERAL)

OTHER RESOURCES

❏ **Focus on Graphing,** SE (GENERAL)

❏ **go.hrw.com**

❏ **www.scilinks.org**

Lesson Plan

Section: Where Did the Elements Come From?

Pacing

Regular Schedule	with lab(s): NA	without lab(s): 1 day
Block Schedule	with lab(s): NA	without lab(s): ½ day

Objectives

1. Describe how the naturally occurring elements form.

2. Explain how a transmutation changes one element into another.

3. Describe how particle accelerators are used to create synthetic elements.

National Science Education Standards Covered

UNIFYING CONCEPTS AND PROCESSES

UCP 1 Systems, order, and organization

UCP 5 Form and function

PHYSICAL SCIENCE—STRUCTURE AND PROPERTIES OF MATTER

PS 2b An element is composed of a single type of atom. When elements are listed in order according to the number of protons (called the atomic number), repeating patterns of physical and chemical properties identify families of elements with similar properties. This "Periodic Table" is a consequence of the repeating pattern of outermost electrons and their permitted energies.

> **KEY**
> **SE** = Student Edition
> **ATE** = Annotated Teacher Edition

Block 5 *(45 minutes)*

FOCUS *5 minutes*

❏ **Bellringer,** ATE (BASIC). Using a list, students locate synthetic elements on the periodic table.

MOTIVATE *5 minutes*

❏ **Identifying Preconceptions,** ATE (BASIC). Discuss with students the meanings of the terms *synthesis* and *synthetic*.

 The Periodic Table

Lesson Plan *continued*

TEACH *25 minutes*

❑ **Transparency,** Nuclear Fusion (GENERAL). This transparency master illustrates one of the fusion reactions thought to take place in the sun: four hydrogen nuclei are fused into one helium nucleus releasing gamma radiation. (Figure 27)

❑ **Transparency,** Nuclear Fusion: Stellar Formation of Carbon-12 (GENERAL). This transparency master illustrates the stellar formation of carbon-12 from beryllium-8. (Figure 28)

❑ **Demonstration,** ATE (GENERAL). This demonstration illustrates how a cloud chamber works.

❑ **Using the Figure,** ATE (GENERAL). Discuss how and why scientists use cloud-chamber tracks like those shown in Figure 29.

CLOSE *10 minutes*

❑ **Reteaching,** ATE (BASIC). Students model nuclear reactions using colored modeling clay and toothpicks.

❑ **Quiz,** ATE (GENERAL). This quiz has students answer questions about the concepts in this lesson.

❑ **Assessment Worksheet: Section Quiz** (GENERAL)

HOMEWORK

❑ **Section Review,** SE (GENERAL). Assign items 1–14.

❑ **Skills Worksheet: Concept Review** (GENERAL)

OTHER RESOURCES

❑ **Skill Builder,** ATE (ADVANCED). This assignment has students prepare reports on the naming of actinides and superheavy elements.

❑ **go.hrw.com**

❑ **www.scilinks.org**

END OF CHAPTER REVIEW AND ASSESSMENT RESOURCES

❑ **Mixed Review,** SE (GENERAL).

❑ **Alternate Assessment,** SE (GENERAL).

❑ **Technology and Learning,** SE (GENERAL).

❑ **Standardized Test Prep,** SE (GENERAL).

❑ **Assessment Worksheet: Chapter Test** (GENERAL)

❑ **Test Item Listing for ExamView® Test Generator**

Name _____ Class _____ Date _____

Skills Practice Lab   **DATASHEETS FOR IN-TEXT LAB**

The Mendeleev Lab of 1869

Russian chemist Dmitri Mendeleev is generally credited as being the first chemist to observe that patterns emerge when the elements are arranged according to their properties. Mendeleev's arrangement of the elements was unique because he left blank spaces for elements that he claimed were undiscovered as of 1869. Mendeleev was so confident that he even predicted the properties of these undiscovered elements. His predictions were eventually proven to be quite accurate, and these new elements fill the spaces that originally were blank in his table.

Use your knowledge of the periodic table to determine the identity of each of the nine unknown elements in this activity. The unknown elements are from the groups in the periodic table that are listed below. Each group listed below contains at least one unknown element.

1   2   11   13   14   17   18

None of the known elements serves as one of the nine unknown elements. No radioactive elements are used during this experiment. The relevant radioactive elements include Fr, Ra, At, and Rn. You may not use your textbook or other reference materials. You have been provided with enough information to determine each of the unknown elements.

OBJECTIVES

Observe the physical properties of common elements.

Observe the properties and trends in the elements on the periodic table.

Draw conclusions and identify unknown elements based on observed trends in properties.

MATERIALS

- blank periodic table
- elemental samples: Ar, C, Sn, Pb
- note cards, 3 in. × 5 in.
- periodic table

Always wear safety goggles and a lab apron to protect your eyes and clothing. If you get a chemical in your eyes, immediately flush the chemical out at the eyewash station while calling to your teacher. Know the location of the emergency lab shower and eyewash station and the procedures for using them.

Name _____ Class _____ Date _____

The Mendeleev Lab of 1869 *continued*

Procedure

1. Put on safety goggles and a lab apron.

2. Use the note cards to copy the information listed on each of the sample cards. If the word *observe* is listed, you will need to visually inspect the sample and then write the observation in the appropriate space.

3. Arrange the note cards of the known elements in a crude representation of the periodic table. In other words, all of the known elements from Group 1 should be arranged in the appropriate order. Arrange all of the other cards accordingly.

4. Once the cards of the known elements are in place, inspect the properties of the unknowns to see where their properties would best "fit" the trends of the elements of each group.

5. Assign the proper element name to each of the unknowns. Add the symbol for each one of the unknown elements to your data table.

6. Clean up your lab station, and return the leftover note cards and samples of the elements to your teacher. Do not pour any of the samples down the drain or in the trash unless your teacher directs you to do so. Wash your hands thoroughly before you leave the lab and after all your work is finished.

TABLE 1 UNKNOWN ELEMENTS IDENTIFICATION

Unknown #	Element	Unknown #	Element
1	Si	6	Sr
2	F	7	Ge
3	Rb	8	Mg
4	Kr	9	Tl
5	Au		

Conclusions

1. Interpreting information Summarize your group's reasoning for the assignment of each unknown. Explain in a few sentences exactly how you predicted the identity of the nine unknown elements.

Answers should include comparisons of properties to elements surrounding

the unknown.

Name _____ Class _____ Date _____

The Mendeleev Lab of 1869 *continued*

Li	
Physical state	solid
Density	0.534 g/cm^3
Hardness	soft, claylike
Conductivity	good
Melting point	180°C
Solubility (H$_2$O)	reacts with water
Color	silver

Ag	
Physical state	solid
Density	10.50 g/cm^3
Hardness	somewhat soft
Conductivity	excellent
Melting point	961°C
Solubility (H$_2$O)	none
Color	silver

Cu	
Physical state	(observe)
Density	8.96 g/cm^3
Hardness	somewhat soft
Conductivity	excellent
Melting point	1083°C
Solubility (H$_2$O)	none
Color	(observe)

C	
Physical state	(observe)
Density	2.10 g/cm^3
Hardness	soft yet brittle
Conductivity	good
Melting point	3550°C
Solubility (H$_2$O)	negligible
Colo	(observe)

Cl$_2$	
Physical state	gas
Density	0.00321 g/cm^3
Hardness	none
Conductivity	very poor
Melting point	−101°C
Solubility (H$_2$O)	slight
Color	greenish yellow

He	
Physical state	gas
Density	0.00018 g/cm^3
Hardness	none
Conductivity	very poor
Melting point	−272°C
Solubility (H$_2$O)	none
Color	colorless

Na	
Physical state	solid
Density	0.971 g/cm^3
Hardness	soft, claylike
Conductivity	good
Melting point	98°C
Solubility (H$_2$O)	reacts rapidly
Color	silver

Ca	
Physical state	solid
Density	1.57 g/cm^3
Hardness	medium
Conductivity	good
Melting point	845°C
Solubility (H$_2$O)	reacts
Color	silvery white

Be	
Physical state	solid
Density	1.85 g/cm^3
Hardness	brittle
Conductivity	excellent
Melting point	1287°C
Solubility (H$_2$O)	none
Color	gray

Sn	
Physical state	(observe)
Density	7.31 g/cm^3
Hardness	somewhat soft
Conductivity	good
Melting point	232°C
Solubility (H$_2$O)	none
Color	(observe)

Ne	
Physical state	gas
Density	0.00090 g/cm^3
Hardness	none
Conductivity	very poor
Melting point	−249°C
Solubility (H$_2$O)	none
Color	colorless

Br$_2$	
Physical state	liquid
Density	3.12 g/cm^3
Hardness	none
Conductivity	very poor
Melting point	−7.2°C
Solubility (H$_2$O)	negligible
Color	reddish brown

K	
Physical state	solid
Density	0.86 g/cm^3
Hardness	soft, claylike
Conductivity	good
Melting point	63°C
Solubility (H$_2$O)	reacts rapidly
Color	silver

Ba	
Physical state	solid
Density	3.6 g/cm^3
Hardness	soft
Conductivity	good
Melting point	710°C
Solubility (H$_2$O)	reacts strongly
Color	silvery white

Xe	
Physical state	gas
Density	0.00585 g/cm^3
Hardness	none
Conductivity	very poor
Melting point	−111.9°C
Solubility (H$_2$O)	none
Color	colorless

Name _____ Class _____ Date _____

The Mendeleev Lab of 1869 *continued*

In	
Physical state	solid
Density	7.31 g/cc
Hardness	very soft
Conductivity	medium
Melting point	157°C
Solubility (H_2O)	none
Color	silvery white

I_2	
Physical state	solid
Density	4.93 g/cm^3
Hardness	soft
Conductivity	very poor
Melting point	113.5°C
Solubility (H_2O)	negligible
Color	bluish-black

Pb	
Physical state	(observe)
Density	11.35 g/cm^3
Hardness	somewhat soft
Conductivity	poor
Melting point	327.5°C
Solubility (H_2O)	none
Color	(observe)

Ar	
Physical state	(observe)
Density	0.00178 g/cm^3
Hardness	none
Conductivity	very poor
Melting point	−189.2°C
Solubility (H_2O)	none
Color	(observe)

Ga	
Physical state	solid
Density	5.904 g/cc
Hardness	soft
Conductivity	medium
Melting point	30°C
Solubility (H_2O)	none
Color	silvery

Cs	
Physical state	solid
Density	1.87 g/cm^3
Hardness	soft
Conductivity	good
Melting point	29°C
Solubility (H_2O)	reacts violently
Color	silvery white

Unknown #1	
Physical state	solid
Density	2.33 g/cm^3
Hardness	brittle
Conductivity	intermediate
Melting point	1410°C
Solubility (H_2O)	none
Color	gray

Unknown #2	
Physical state	gas
Density	0.00170 g/cm^3
Hardness	none
Conductivity	very poor
Melting point	−219.6°C
Solubility (H_2O)	slight
Color	pale yellow

Unknown #3	
Physical state	solid
Density	1.53 g/cm^3
Hardness	soft
Conductivity	good
Melting point	39°C
Solubility (H_2O)	reacts violently
Color	silvery white

Unknown #4	
Physical state	gas
Density	0.00374 g/cm^3
Hardness	none
Conductivity	very poor
Melting point	−156.6°C
Solubility (H_2O)	none
Color	colorless

Unknown #5	
Physical state	solid
Density	19.3 g/cm^3
Hardness	soft
Conductivity	excellent
Melting point	1064°C
Solubility (H_2O)	none
Color	gold

Unknown #6	
Physical state	solid
Density	2.54 g/cm^3
Hardness	somewhat soft
Conductivity	good
Melting point	769°C
Solubility (H_2O)	reacts rapidly
Color	silvery white

Unknown #7	
Physical state	solid
Density	5.32 g/cm^3
Hardness	fairly brittle
Conductivity	fair to poor
Melting point	937°C
Solubility (H_2O)	none
Color	gray

Unknown #8	
Physical state	solid
Density	1.74 g/cm^3
Hardness	medium
Conductivity	good
Melting point	651°C
Solubility (H_2O)	reacts slowly
Color	silvery white

Unknown #9	
Physical state	solid
Density	11.85 g/cc
Hardness	very soft
Conductivity	medium
Melting point	303°C
Solubility (H_2O)	none
Color	silvery white

Reactivity of Halide Ions

Teacher Notes

TIME REQUIRED One 45-minute lab period

SKILLS ACQUIRED
Collecting data
Identifying patterns
Inferring
Interpreting
Organizing and analyzing data

RATING

Easy ← 1 2 3 4 → Hard

Teacher Prep–3
Student Set-Up–2
Concept Level–2
Clean Up–3

THE SCIENTIFIC METHOD

Make Observations Students observe chemical reactions.

Analyze the Results Analysis question 1 requires analysis of the results.

Draw Conclusions Conclusions questions 1, 2, and 3 asks students to draw conclusions from their data.

Communicate the Results Conclusions questions 2 and 3.

MATERIALS

Materials for this lab activity can be purchased from WARD'S. See the *Master Materials List* on the **One-Stop Planner** CD-ROM for ordering instructions.

To prepare 50 mL of 0.1 M $AgNO_3$ solution, dissolve 0.85 g of $AgNO_3$ in enough distilled water to make 50 mL of solution.

To prepare 50 mL of 0.5 M $Ca(NO_3)_2$ solution, dissolve 4.1 g of $Ca(NO_3)_2$ in enough distilled water to make 50 mL of solution.

To prepare 50 mL of 0.2 M KBr solution, dissolve 1.2 g of KBr in enough distilled water to make 50 mL of solution.

To prepare 125 mL of 0.2 M KI, dissolve 4.1 g of KI in enough distilled water to make 125 mL of solution.

To prepare 100 mL of 0.1 M NaCl solution, dissolve 0.6 g of NaCl in enough distilled water to make 100 mL of solution.

To prepare 125 mL of 0.1 M NaF solution, dissolve 0.5 g of NaF in enough distilled water to make 125 mL of solution.

To prepare 100 mL of 0.2 M $Na_2S_2O_3$ solution, dissolve 5.0 g of $Na_2S_2O_3 \cdot 5H_2O$ in enough distilled water to make 100 mL of solution.

Reactivity of Halide Ions *continued*

To prepare 500 mL of 4 M aqueous NH_3, dilute 134 mL of concentrated ammonia water to a volume of 500 mL.

To prepare a 3% starch solution, dissolve 3.0 g of soluble starch in approximately 80 mL of water. Boil, and then dilute to 100 mL of solution.

If a mixture of two ions is used for an unknown, double the concentration of each original solution.

SAFETY CAUTIONS

Safety goggles, gloves, and a lab apron must be worn at all times. Read all safety precautions, and discuss them with your students.

Remind students of the following safety precautions:

• Always wear safety goggles, gloves, and a lab apron to protect your eyes, hands, and clothing. If you get a chemical in your eyes, immediately flush the chemical out at the eyewash station while calling to your teacher. Know the location of the emergency lab shower and eyewash station and the procedure for using them.

• Do not touch any chemicals. If you get a chemical on your skin or clothing, wash the chemical off at the sink while calling to your teacher. Make sure you carefully read the labels and follow the precautions on all containers of chemicals that you use. If there are no precautions stated on the label, ask your teacher what precautions you should follow. Do not taste any chemicals or items used in the laboratory. Never return leftovers to their original containers; take only small amounts to avoid wasting supplies.

• Call your teacher in the event of a spill. Spills should be cleaned up promptly, according to your teacher's directions.

• Never put broken glass into a regular waste container. Broken glass should be disposed of properly in the broken-glass waste container.

• In case of a spill, use a dampened cloth or paper towel (or more than one towel if necessary) to mop up the spill. Then rinse the towel in running water at the sink, wring it out until it is only damp, and put it in the trash. In the event of an ammonia spill, dilute the spill with water and then proceed as described.

DISPOSAL

Provide a plastic dishpan to collect the rinsings for step 10. Combine all solutions. Add enough 0.2 M KI to precipitate all of the Ag^+ as AgI. Then dilute 100-fold and pour the mixture down the drain.

TECHNIQUES TO DEMONSTRATE

Review proper safety precautions required when dealing with chemicals.

Show the students the proper method for filling and using a pipet.

Review proper disposal and clean up procedures with students.

Name _____ Class _____ Date _____

Skills Practice Lab

MICROSCALE LAB

Reactivity of Halide Ions

The four halide salts used in this experiment are found in your body. Although sodium fluoride is poisonous, trace amounts seem to be beneficial to humans in the prevention of tooth decay. Sodium chloride is added to most of our food to increase flavor while masking sourness and bitterness. Sodium chloride is essential for many life processes, but excessive intake appears to be linked to high blood pressure. Sodium bromide is distributed throughout body tissues, and in the past it has been used as a sedative. Sodium iodide is necessary for the proper operation of the thyroid gland, which controls cell growth. The concentration of sodium iodide is almost 20 times greater in the thyroid than in blood. The need for this halide salt is the reason that about 10 ppm of NaI is added to packages of table salt labeled "iodized."

The principal oxidation number of the halogens is -1. However, all halogens except fluorine may have other oxidation numbers. The specific tests you will develop in this experiment involve the production of recognizable precipitates and complex ions. You will use your observations to determine the halide ion present in an unknown solution.

MATERIALS

- 24-well microplate
- $AgNO_3$, 0.1 M
- $Ca(NO_3)_2$, 0.5 M
- gloves
- KBr, 0.2 M
- KI, 0.2 M
- lab apron
- $Na_2S_2O_3$, 0.2 M
- NaCl, 0.1 M
- NaF, 0.1 M
- NaOCl (commercial bleach), 5%
- $NH_3(aq)$, 4 M
- safety goggles
- starch solution, 3%
- thin-stemmed pipets (12)

Always wear safety goggles and a lab apron to protect your eyes and clothing. If you get a chemical in your eyes, immediately flush the chemical out at the eyewash station while calling to your teacher. Know the location of the emergency lab shower and eyewash station and the procedures for using them.

Do not touch any chemicals. If you get a chemical on your skin or clothing, wash the chemical off at the sink while calling to your teacher. Make sure you carefully read the labels and follow the precautions on all containers of chemicals that you use. If there are no precautions stated on the label, ask your teacher what precautions to follow. Do not taste any chemicals or items used in the laboratory. Never return leftovers to their original container; take only small amounts to avoid wasting supplies.

Call your teacher in the event of a spill. Spills should be cleaned up promptly, according to your teacher's directions.

Name _____ Class _____ Date _____

Reactivity of Halide Ions *continued*

Acids and bases are corrosive. If an acid or base spills onto your skin or clothing, wash the area immediately with running water. Call your teacher in the event of an acid spill. Acid or base spills should be cleaned up promptly.

 Never put broken glass in a regular waste container. Broken glass should be disposed of separately according to your teacher's instructions.
   Never stir with a thermometer because the glass around the bulb is fragile and might break.

OBJECTIVES

Observe the reactions of the halide ions with different reagents.

Analyze data to determine characteristic reactions of each halide ion.

Infer the identity of unknown solutions.

Procedure

1. Put on safety goggles, gloves, and a lab apron.

2. Put 5 drops of 0.1 M NaF into each of four wells in row A, as shown in **Figure 1.** Put 5 drops of 0.1 M NaCl into each of the wells in row B. Put 5 drops of 0.2 M KBr into each of the wells in row C and 5 drops of 0.2 M KI into each of the wells in Row D. Reserve rows E and F for unknown solutions.

Figure 1

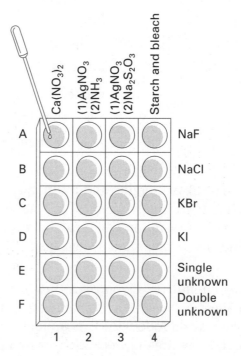

3. Add 5 drops of 0.5 M $Ca(NO_3)_2$ solution to each of the four halide solutions in column 1. Record your observations in the **Table 1.**

4. Add 2 drops of 0.1 M $AgNO_3$ solution to each of the halides in columns 2 and 3. Record in **Table 1** the colors of the precipitates formed.

5. Add 5 drops of 4 M $NH_3(aq)$ to the precipitates in column 2. Record your observations in the **Table 1.**

6. Add 5 drops of 0.2 M $Na_2S_2O_3$ solution to the precipitates in column 3. Record your observations in **Table 1.**

7. To the halides in column 4, add 5 drops of starch solution and 1 drop of 5% bleach solution. Record your observations. Save the results of testing the known halide solutions for comparison with the tests of the unknown solutions.

8. Obtain an unknown solution. Put 5 drops of the unknown in each of the four wells in row E. Add the reagents to each well as you did in steps 3–6. Compare the results with those of the known halides in rows A–D. Record your findings in **Table 1,** and identify the unknown.

9. Obtain an unknown solution containing a mixture of two halide ions. Place 5 drops of the unknown mixture in each of the four wells in row F. Add the reagents to each well as you did in steps 3–6. Record your results. Compare the results with those of the known halides in rows A–D. Identify the halides in the double unknown solution.

10. Rinse the microplate into a trough or dishpan provided by your teacher. Clean all apparatus and your lab station. Return equipment to its proper place. Dispose of chemicals and solutions in the containers designated by your teacher. Do not pour any chemicals down the drain or in the trash unless your teacher directs you to do so. Wash your hands thoroughly before you leave the lab and after all work is finished.

TABLE 1: RESULTS OF THE REACTIONS OF HALIDE SALTS

Halide salts	$Ca(NO_3)_2$	$AgNO_3$	$AgNO_3$ + NH_3	$AgNO_3$ + $Na_2S_2O_3$	NaOCl + starch
NaF	clear	clear	clear	clear	clear
NaCl	clear	white ppt	dissolves completely	dissolves completely	clear
KBr	clear	pale yellow ppt	dissolves completely	dissolves completely	yellow tinge
KI	clear	bright yellow ppt	bright yellow ppt	bright yellow ppt	indigo ppt
Single unknown	clear	pale yellow ppt	dissolves completely	dissolves completely	yellow tinge
Double unknown	clear	pale yellow ppt	bright yellow ppt	bright yellow	indigo ppt

Name _____ Class _____ Date _____

Reactivity of Halide Ions *continued*

Analysis

1. **Analyzing Data** Which procedure(s) confirm(s) the presence of (a) F^- ions, (b) Cl^- ions, (c) Br^- ions, (d) I^- ions?

   a. Procedure 3

   b. Procedures 4 and 5

   c. Procedures 4 and 6

   d. Procedure 7

Conclusions

1. **Drawing Conclusions** What generalizations can be made about silver halides?

   The silver halides produced in the reactions are all insoluble, except for AgF.

   _____

2. **Applying Conclusions** In nuclear explosions or accidents, iodine-131, a radioactive fission product, can become dispersed in the atmosphere. Eventually, the iodine isotope will fall onto the ground and be absorbed by plants. Explain how radiation from iodine-131 could become concentrated in the human body and cause a growth disorder.

   Cattle eat the plants and concentrate the radioactive iodine in their milk. If

   humans drink the milk, the radioactive iodine will concentrate in their

   thyroid gland, causing a possible growth disorder.

   _____

3. **Defending Conclusions** Identify your unknown(s) and use your experimental evidence to support your identifications.

   Students' answers will vary.

   _____

   _____

Skills Practice Lab

Exploring the Periodic Table

Time Required

One lab period

Teacher's Notes

Graphing Calculator and Sensors

TIPS AND TRICKS

- Each lab team must have the Periodic application installed on its graphing calculator. This application can be obtained by downloading from the TI website.

TECHNIQUES TO DEMONSTRATE

Instruct students how to navigate through menus within the Periodic application. Menu items at the bottom of the screen are accessed by pressing the calculator key located directly below. Items in vertical menus can be selected by moving the cursor up or down with the direction keys and then pressing ENTER.

Answers to Problems

1. Atomic mass

   a. bromine, Br = 79.904 amu

   b. tungsten, W = 183.84 amu

   c. potassium chloride, KCl = 74.551 amu

   d. glucose, $C_6H_{12}O_6$ = 180.15768 amu

2. The electronegativity increases as you move from the left to the right of the periodic table.

3. Iridium has the greatest density at 22.65 g/cm^3.

4. Density

   a. argon, Ar = 0.0017824 g/cm^3

   b. aluminum, Al = 2.70 g/cm^3

   c. strontium, Sr = 2.64 g/cm^3

   d. iron, Fe = 7.86 g/cm^3

5. The alkali earth metals are beryllium, magnesium, calcium, strontium, barium, and radium.

Name _____ Class _____ Date _____

Skills Practice Lab

GRAPHING CALCULATOR LAB

Exploring the Periodic Table

The periodic table of elements, found in your chemistry textbook, contains a wealth of information. Within the periodic table, each of the known elements is listed by name, as well as symbol. Other information found in the periodic table includes the atomic weight along with the atomic number of each element. Each element in the periodic table is arranged according to its properties. For example, the elements listed in the first column of the table are the alkali metals. These elements all have low densities, low melting points, and good electrical conductivity.

The same wealth of information found in the periodic table can be found within the Periodic application on your TI-83 Plus graphing calculator. In this activity, you will learn how to navigate through the Periodic application and access this information. In addition to the information found within the periodic table, the Periodic application contains additional information about each element such as density, melting point, boiling point, electronegativity, and oxidation states.

OBJECTIVES

- **Explore** the periodic table of elements using the Periodic application.

- **Identify** various elements found in the periodic table.

- **Observe** elemental trends using the graphing feature of the Periodic application.

- **Calculate** the atomic weight of various chemical compounds.

EQUIPMENT

- TI-83 Plus graphing calculator

- Periodic application

PRELAB

Navigational menus for the Periodic application can be found at the bottom of the screen.

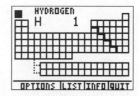

To select one of the menu items found at the bottom of the screen, press the calculator key located directly below that item. For example, to select LIST press ZOOM, and to select INFO press TRACE. To select OPTIONS, press either Y= or WINDOW because it has two calculator keys located below it.

To select items from a vertical screen, such as the one shown below, move the highlight cursor up or down through the list by pressing ▲ or ▼. When the option you wish to select is highlighted, press ᴇɴᴛᴇʀ or select OK.

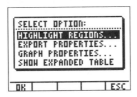

Procedure

1. Turn on the calculator. Press APPS, and then press the calculator key for the *number* that precedes the PERIODIC application. You are now at the main screen of the program.

TABLE SCREEN

When the Periodic application is started, the table screen is displayed. Follow the instructions below to move through the table and select the element sodium.

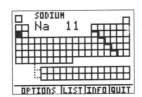

2. Move the cursor from cell to cell by pressing ▲, ▼, ◄, or ►. Position the cursor over the cell for sodium. When the cursor is positioned properly, the name of the element is displayed at the top of the screen.

3. Press ᴇɴᴛᴇʀ to display detailed properties for the element sodium.

4. Press ▲ or ▼ to scroll through the properties for sodium. To scroll the screen a page at a time, press ᴀʟᴘʜᴀ and then ▲ or ▼.

When the properties screen is displayed, values for each of the properties can be copied to the calculator Home screen for use in calculations. Follow the instructions below to calculate the atomic mass of NaCl.

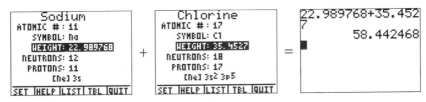

5. Press ▼ to move the cursor down until WEIGHT is highlighted.

6. Select SET to copy the value for atomic mass to the calculator Home screen.

7. Select QUIT to exit the Periodic application and press ⎣ + ⎦.

8. Restart the Periodic application.

9. Repeat Steps 2–4 to select the element chlorine.

Name _____ Class _____ Date _____

Exploring the Periodic Table *continued*

10. Repeat Steps 5–6 to copy the atomic mass of chlorine to the calculator Home screen.

11. Select QUIT to exit the Periodic application. Press [ENTER] to calculate the atomic mass of NaCl.

DISPLAYING ELEMENT GROUPS

Groups of elements can be highlighted and displayed on the table screen. Follow the instructions below to highlight the alkali metals on the table screen.

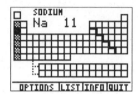

1. From the table screen, select OPTIONS to display the Options menu.

2. Press [ENTER] to select HIGHLIGHT REGIONS from the Options menu.

3. Press [▼] to move the cursor down until ALKALI METALS is highlighted, and press [ENTER]. The table screen will now be displayed with the selected area highlighted.

GRAPHING ELEMENTS BY PROPERTIES

To help identify trends within the periodic table, the Periodic application has five graphing options. Each graph has the atomic number displayed on the x-axis and the atomic radius, 1st ionization energy, electronegativity, density, or melting point displayed on the y-axis. Follow the instructions below to display a graph of atomic number versus density.

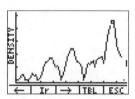

1. From the table screen, select OPTIONS to display the Options menu.

2. Press [▼] to move the cursor down until GRAPH PROPERTIES is highlighted, and press [ENTER].

3. Press [▼] to move the cursor down until DENSITY is highlighted, and press [ENTER]. A graph of density versus atomic number will be displayed.

4. Press [◄] or [►] to move the trace cursor from element to element.

Name _____ Class _____ Date _____

Exploring the Periodic Table *continued*

DISPLAYING ELEMENTS AS A LIST

Instead of the table screen, the elements can be displayed in list format. Once in a list, the elements can be sorted by atomic number, name, or symbol. Follow the instructions below to display a list of the elements sorted by symbol.

```
Ac   ACTINIUM      89
Ag   SILVER        47
Al   ALUMINIUM     13
Am   AMERICIUM     95
Ar   ARGON         18
As   ARSENIC       33
At   ASTATINE      85
          SORT TBL QUIT
```

1. From the table screen, select LIST to display a list of the elements.

2. Select SORT from the list screen.

3. Press ▼ to move the cursor down until SYMBOL is highlighted, and press ENTER. The list of elements will now be sorted according to symbol.

4. To select an element, press ▲ or ▼ to move the cursor from element to element. When your element is highlighted, press ENTER to display properties for that element.

SAMPLE PROBLEMS

1. Using the Periodic application, calculate the atomic mass for each of the following.

   a. Bromine, Br

   b. Tungsten, W

   c. Potassium chloride, KCl

   d. Glucose, $C_6H_{12}O_6$

2. Display the graph of electronegativity versus atomic number. Using the graph, determine what happens to the electronegativity as you move from left to right across the periodic table.

3. Display the graph of density versus atomic number. Using the graph, determine which element has the greatest density. Identify the element, and list its density.

4. Determine the density for each of the following.

   a. argon, Ar

   b. aluminum, Al

   c. strontium, Sr

   d. iron, Fe

5. Using the Highlight Region option, identify the elements that make up the alkali earth metals.

Answer Key

Concept Review: How Are Elements Organized?

1. Li, Na, K, Rb, Cs, and Fr all have a single electron in their outer shells; this single electron is transferred to Cl and its seven valence electrons to form a stable octet. The products have similar properties because they have similar electron configurations.
2. Newlands arranged the elements in groups resulting in a table with eight columns. He discovered that the elements in each column had similar properties.
3. Mendeleev arranged the elements in eight columns in order of increasing atomic mass.
4. Mendeleev had gaps in his table that he surmised were undiscovered elements.
5. Moseley used X-ray spectra and found that his spectral lines correlated to increasing atomic number rather than atomic mass. The discrepancies in arrangement by atomic mass disappeared table when the elements were arranged in order of increasing atomic number.
6. Mendeleev did not have the technology available to him that Moseley did.
7. An atom's valence electrons, or outermost electrons, participate in chemical reactions with other atoms, so elements with the same number of valence electrons tend to react in similar ways.
8. When elements are arranged according to increasing atomic number, elements with the same number of valence electrons tend to occur at periodic intervals. Because they have the same number of valence electrons, they have similar properties.
9. Each member of a group has the same number of electrons in its outer shell.
10. A group is a vertical column on the periodic table, and a period is a horizontal row.

Concept Review: Tour of the Periodic Table

1. alkali metals
2. main group elements
3. alkaline earth metals
4. transition metals
5. halogens
6. noble gases
7. noble gases
8. halogens
9. transition metals
10. alkali metals
11. metals
12. hydrogen
13. Groups 1, 2, 13, 14, 15, 16, 17, 18
14. The electron configurations of main-groups elements are regular and consistent.
15. Group 2 must lose two electrons and Group 1 must lose one electron to achieve the noble gas configuration. Because losing two electrons requires about twice the energy required to lose one electron, Group 2 elements are less reactive.
16. Helium is unreactive and will not cause an undesirable combustion reaction. The fuel-to-oxygen ratio is also carefully controlled to achieve the best result in the welding process.
17. The halogens need one electron to achieve the noble gas configuration. The alkali metals have one electron that is easily removed; therefore, they combine readily in a 1:1 ratio to form a salt.
18. Iron alloys, such as steel, are harder, stronger, and more resistant to corrosion than pure iron.
19. metals
20. 3, 12
21. d, d-block
22. non-metals, 13, 16
23. lanthanides
24. actinides
25. actinides

Concept Review: Trends in the Periodic Table

1. ionization energy
2. bond radius
3. electron affinity
4. electronegativity
5. increases, decreasing
6. d
7. a
8. c
9. The electron cloud model is based on the probability of finding an electron at a specific location. As you move farther out from the nucleus, the probability of finding electrons becomes less and less. With this model there is not a well-defined boundary of the individual atom.
10. Na; because it has one more energy level than Li.
11. Electron shielding is the reduction of the attractive force between a positively charged nucleus and its outermost electrons due to the cancellation of some of the positive charge by the negative charge of the other electrons.
12. As the outermost electrons are pulled closer to the nucleus, they also get closer to one another and repulsion gets stronger. At Group 13, the electrons will not come closer to the nucleus because the electrons repel each other.
13. Each element has one more occupied energy levels than the one above it. Therefore, the outermost electrons are farther from the nucleus as you move down a group. Also, each successive element contains more electrons between the nucleus and the outermost electrons. These innermost electrons shield the outermost electrons from the full attractive force of the nucleus, thereby making it easier to remove valence electrons.

14. As you more across a period, each atom has one more proton and one more electron in the same principal energy level as the one before it. Therefore, because electron shielding does not change, the nuclear charge increases across a period, attracting the electrons more strongly.
15.

	General Trend	
	Across a Period	Down a Group
Ionization Energy	increases	decreases
Atomic Radius	decreases	increases
Electronegativity	increases	decreases
Ionic Size	decreases	increases
Electron affinity	increases	decreases

Concept Review: Where Did the Elements Come From?

1. carbon, hydrogen, oxygen, nitrogen, phosphorous, and sulfur
2. energy
3. matter, electrons, protons, and neutrons
4. hydrogen, helium
5. helium
6. fusion reactions
7. supernova
8. transmutation
9. synthetic
10. particle accelerators
11. The 93 naturally occurring elements are found on Earth or on stars. The remaining 20 elements are synthetic.
12. Cyclotrons cannot accelerate particles fast enough because as the particles accelerate, they become more massive, making it increasingly difficult to achieve further acceleration.
13. A synchrotron times its energy pulses to match the acceleration of the particle, thereby accelerating particles to enormous speeds.
14. Only a few atoms are created and they last for tiny fractions of a second.

Answer Key

Quiz–Section: How Are the Elements Organized?

1. d	**6.** b
2. a	**7.** b
3. d	**8.** b
4. a	**9.** b
5. a	**10.** d

Quiz–Section: Tour of the Periodic Table

1. b	**6.** d
2. a	**7.** c
3. b	**8.** c
4. b	**9.** c
5. b	**10.** d

Quiz–Section: Trends in the Periodic Table

1. c	**6.** c
2. c	**7.** a
3. a	**8.** c
4. b	**9.** a
5. a	**10.** d

Quiz–Section: Where Did the Elements Come From?

1. d	**6.** d
2. c	**7.** c
3. a	**8.** c
4. b	**9.** d
5. d	**10.** d

Chapter Test

1. b	**11.** b
2. d	**12.** a
3. b	**13.** d
4. b	**14.** c
5. d	**15.** a
6. a	**16.** c
7. b	**17.** d
8. b	**18.** a
9. b	**19.** d
10. b	**20.** c

21. Noble gases are not reactive because their outermost shells are completely filled with electrons, which makes these elements very stable.
22. In a cyclotron, charged particles are given successive pulses of energy that accelerate them. These energetic particles collide with atomic nuclei and combine to produce nuclei of a higher atomic number.
23. The star will collapse and then explode to form a red giant star. The fusion in this process creates elements with atomic numbers greater than those of iron and nickel. If the star forms a super nova, the elements formed in the star will be scattered across space.
24. In decreasing order: As, Sb, Sn, In, Rb
25. In increasing order: I, Sb, In, Sr, Rb

The Periodic Table

MULTIPLE CHOICE

1. The idea of arranging the elements in a table according to their chemical and physical properties is attributed to
   a. Newlands.
   b. Moseley.
   c. Bohr.
   d. Ramsay.

   Answer: A   Difficulty: I   Section: 1   Objective: 1

2. From gaps in his table, Mendeleev predicted the existence of several elements and their
   a. atomic numbers.
   b. colors.
   c. properties.
   d. radioactivity.

   Answer: C   Difficulty: I   Section: 1   Objective: 1

3. Mendeleev is credited with developing the first successful
   a. periodic table.
   b. method for determining atomic number.
   c. test for radioactivity.
   d. use of X rays.

   Answer: A   Difficulty: I   Section: 1   Objective: 1

4. The principle that states that the physical and chemical properties of the elements are periodic functions of their atomic numbers is
   a. the periodic table.
   b. the periodic law.
   c. the law of properties.
   d. Mendeleev's law.

   Answer: B   Difficulty: I   Section: 1   Objective: 2

5. The periodic law states that

   a. no two electrons with the same spin can be found in the same place in an atom.
   b. the physical and chemical properties of the elements are functions of their atomic numbers.
   c. electrons exhibit properties of both particles and waves.
   d. the chemical properties of elements can be grouped according to periodicity, but physical properties cannot.

   Answer: B   Difficulty: I   Section: 1   Objective: 2

6. Elements in a group or column in the periodic table can be expected to have similar
   a. atomic masses.
   b. atomic numbers.
   c. numbers of neutrons.
   d. properties.

   Answer: D   Difficulty: I   Section: 1   Objective: 2

7. The atomic number of sodium, the first element in Period 3, is 11. The atomic number of the second element in this period is
   a. 3.
   b. 10.
   c. 12.
   d. 18.

   Answer: C   Difficulty: I   Section: 1   Objective: 2

8. Elements in a group have similar
   a. reactivities.
   b. densities.
   c. symbols.
   d. electron configurations.

   Answer: D   Difficulty: I   Section: 1   Objective: 2

9. An element that has an electron configuration of $[He]2s^22p^3$ is in Period _____ of the periodic table.
   a. 1
   b. 2
   c. 3
   d. 4

   Answer: B   Difficulty: II   Section: 1   Objective: 2

10. An element that has an electron configuration of $[Ne]3s^23p^3$ is in Group _____ of the periodic table.
    a. 2
    b. 3
    c. 6
    d. 15
    Answer: D   Difficulty: II   Section: 1   Objective: 2

11. To which group of the periodic table do lithium and potassium belong?
    a. alkali metals
    b. transition metals
    c. halogens
    d. noble gases
    Answer: A   Difficulty: I   Section: 2   Objective: 1

12. To which group of the periodic table do fluorine and chlorine belong?
    a. alkaline-earth metals
    b. transition elements
    c. halogens
    d. actinides
    Answer: C   Difficulty: I   Section: 2   Objective: 1

13. The outer electron configuration of an alkali metal has
    a. 1 electron in the *s* orbital.
    b. 2 electrons in the *s* orbital.
    c. 1 electron in the *p* orbital.
    d. 2 electrons in the *p* orbital.
    Answer: A   Difficulty: I   Section: 2   Objective: 1

14. The elements in Group 17 are known by what name?
    a. halogens
    b. alkali metals
    c. alkaline-earth metals
    d. noble gases
    Answer: A   Difficulty: I   Section: 2   Objective: 1

15. Transition elements are found in which of the following group(s) of the periodic table?
    a. Group 1
    b. Groups 1 and 2
    c. Groups 3–12
    d. Group 18
    Answer: C   Difficulty: I   Section: 2   Objective: 2

16. An element that has four electrons in its outermost *d* orbitals and one electron in its outermost *s* orbital is a member of what group in the periodic table?
    a. Group 1
    b. Groups 4
    c. Group 5
    d. Group 15
    Answer: C   Difficulty: II   Section: 2   Objective: 2

17. Moving from left to right across _____, electrons are being added to the 4*f* orbitals.
    a. the noble gases
    b. the halogens
    c. the lanthanides
    d. Period 4
    Answer: C   Difficulty: I   Section: 2   Objective: 2

18. A property of all actinides is that they are
    a. members of Period 7.
    b. radioactive.
    c. nonmetals.
    d. Both (a) and (b)
    Answer: D   Difficulty: II   Section: 2   Objective: 2

19. A metal is called malleable if it
    a. has a shiny appearance.
    b. can be hammered into sheets.
    c. can be squeezed out into a wire.
    d. exists naturally as an element.
    Answer: B   Difficulty: I   Section: 2   Objective: 2

20. A solution of two or more metals is
    a. an insulator.
    b. a jelly.
    c. brittle.
    d. an alloy.
    Answer: D   Difficulty: I   Section: 2   Objective: 2

21. Trends in the properties of elements in a group or period can be explained in terms of
    a. binding energy.
    b. atomic number.
    c. electron configuration.
    d. electron affinity.

    Answer: C   Difficulty: I   Section: 3   Objective: 1

22. The effect of inner electrons on the attraction between the nucleus and the outer electrons of an atom is called
    a. electron affinity.
    b. electron ionization.
    c. electron shielding.
    d. electronegativity.

    Answer: C   Difficulty: I   Section: 3   Objective: 1

23. Trends in the periodic table indicate that the element with the greatest ionization energy is in which of the following periods and groups?
    a. Period 2, Group 1
    b. Period 7, Group 2
    c. Period 1, Group 18
    d. Period 6, Group 17

    Answer: C   Difficulty: I   Section: 3   Objective: 1

24. One method of measuring the size of an atom involves calculating a value that is _____ the distance between the nuclei of two bonded atoms.
    a. twice
    b. equal to
    c. half
    d. one-quarter

    Answer: C   Difficulty: I   Section: 3   Objective: 2

25. Going down a group in the periodic table, electron shielding generally causes the effective nuclear charge to
    a. increase.
    b. remain the same.
    c. decrease.
    d. vary unpredictably.

    Answer: C   Difficulty: I   Section: 3   Objective: 2

26. Going across a period in the periodic table, electron shielding generally has little effect. As a result, the effective nuclear charge
    a. increases.
    b. remain the same.
    c. decreases.
    d. varies unpredictably.

    Answer: A   Difficulty: II   Section: 3   Objective: 2

27. Which is the best reason that the atomic radius generally increases with atomic number in each group of elements?
    a. The nuclear charge increases.
    b. The number of neutrons increases.
    c. The number of energy levels increases.
    d. A new octet forms.

    Answer: C   Difficulty: II   Section: 3   Objective: 2

28. For the alkaline-earth metals, atoms with the smallest radii have the
    a. largest atomic numbers.
    b. greatest volumes.
    c. most mass.
    d. highest ionization energies.

    Answer: D   Difficulty: II   Section: 3   Objective: 2

29. Trends in the periodic table indicate that an element in which of the following periods and groups will have the smallest anion (negative ion) radius?
    a. Period 2, Group 1
    b. Period 7, Group 2
    c. Period 4, Group 16
    d. Period 1, Group 17

    Answer: D   Difficulty: I   Section: 3   Objective: 4

30. Trends in the periodic table indicate that an element in which of the following periods and groups will have the largest cation (positive ion) radius?
   a. Period 7, Group 1
   b. Period 2, Group 2
   c. Period 4, Group 2
   d. Period 1, Group 17
   Answer: A   Difficulty: I   Section: 3   Objective: 4

31. Electron affinity tends to
   a. decrease across a period and decrease down a group.
   b. increase across a period and increase down a group.
   c. decrease across a period and increase down a group.
   d. increase across a period and decrease down a group.
   Answer: D   Difficulty: II   Section: 3   Objective: 4

32. The process that changes one element into another different element is called
   a. transfiguration.
   b. transformation.
   c. transmutation.
   d. transgeneration.
   Answer: C   Difficulty: I   Section: 4   Objective: 2

COMPLETION

33. Newlands called the repetitive pattern of the properties of the elements in his table the *law of octaves* because the pattern repeated every _____ elements.
   Answer: eight   Difficulty: I   Section: 1   Objective: 1

32. The development of the first successful periodic table is credited to _____.
   Answer: Mendeleev Difficulty: I   Section: 1   Objective: 1

33. That physical and chemical properties of the elements are functions of their atomic numbers is a statement of the _____.
   Answer: periodic law
   Difficulty: I   Section: 1   Objective: 1

34. Elements in a(n) _____ in the periodic table can be expected to have similar properties.
   Answer: group or column
   Difficulty: I   Section: 1   Objective: 2

35. A period is a(n) _____ of blocks in the periodic table.
   Answer: horizontal row Difficulty: I   Section: 1   Objective: 2

36. An element that has an electron configuration of $[He]2s^22p^3$ is in Period _____, Group _____
   Answer: Period 2, Group 15
   Difficulty: II   Section: 1   Objective: 2

37. An electron that is found in the outermost shell of an atom and that determines the atom's chemical properties is called a _____ electron.
   Answer: valence   Difficulty: I   Section: 1   Objective: 2

38. The alkaline-earth metals are found in the periodic table in Group _____.
   Answer: 2   Difficulty: I   Section: 2   Objective: 1

39. Elements in Group 17 of the periodic table are called the _____.
   Answer: halogens   Difficulty: I   Section: 2   Objective: 1

40. Elements whose electron configuration ends in *s* and *p* orbitals are called the _____ elements.
   Answer: main-group Difficulty: I   Section: 2   Objective: 1

41. The only element that does not belong to a group is _____.
    Answer: hydrogen   Difficulty: I   Section: 2   Objective: 1

42. Except for helium, a noble gas has a total of _____ electrons in its outermost *s* and *p* orbitals.
    Answer: 8   Difficulty: II   Section: 2   Objective: 1

43. What is the number of the group in the periodic table that contains an element with an electron configuration of $[Xe]4f^{14}5d^96s^1$?
    Answer: Group 10   Difficulty: II   Section: 2   Objective: 2

44. All metals are good conductors of _____.
    Answer: electricity   Difficulty: I   Section: 2   Objective: 2

45. A member of the rare-earth series of elements, whose atomic numbers range from 58 to 71 is classified as a(n) _____.
    Answer: lanthanide   Difficulty: I   Section: 2   Objective: 2

46. All elements in the _____ series are radioactive.
    Answer: actinide   Difficulty: II   Section: 2   Objective: 2

47. A metal can be squeezed out into a wire because it is _____.
    Answer: ductile   Difficulty: I   Section: 2   Objective: 2

48. The removal of an electron from a neutral atom, A, can be described as:
    A + ionization _____ $\rightarrow A^+ + e^-$
    Answer: energy   Difficulty: I   Section: 3   Objective: 1

49. The effect that causes the outermost electrons in an atom or ion to be held less tightly to the nucleus because of inner electrons is called _____.
    Answer: electron shielding
    Difficulty: I   Section: 3   Objective: 1

50. A measure of the ability of an atom in a chemical compound to attract electrons is called
    _____.
    Answer: electronegativity
    Difficulty: I   Section: 3   Objective: 3

51. A neutral atom's electron affinity is defined as its change in _____ as it gains an electron.
    Answer: energy   Difficulty: I   Section: 3   Objective: 4

52. The currently most-accepted scientific model of the universe's beginnings is an explosion called the _____.
    Answer: big bang   Difficulty: I   Section: 4   Objective: 1

53. In the following nuclear reaction that takes place in the sun,
    $$4^1_1 H \text{ nuclei} \rightarrow {}^4_2He \text{ nucleus} + \gamma \text{ radiation}$$
    two protons change into two _____.
    Answer: neutrons   Difficulty: II   Section: 4   Objective: 1

54. All elements in the periodic table with atomic numbers greater than that of the element _____ are formed by the explosions of collapsing stars.
    Answer: iron   Difficulty: II   Section: 4   Objective: 1

55. In the sun, the fusion of two helium nuclei produces a nucleus of the element _____ and the release of energy in the form of gamma radiation.
    Answer: beryllium/Be   Difficulty: I   Section: 4   Objective: 1

56. Transmutation reactions can occur when the _____ of various elements are bombarded by alpha particles.

Answer: nuclei   Difficulty: I   Section: 4   Objective: 2

57. The production of synthetic elements is an example of a type of nuclear reaction that is called a(n) _____ .

Answer: transmutation Difficulty: I   Section: 4   Objective: 2

58. Cyclotrons and synchrotrons accelerate charged particles by repeatedly supplying them with pulses of _____.

Answer: energy   Difficulty: I   Section: 4   Objective: 3

59. A synchrotron can accelerate charged particles to high speeds because it compensates for the increase in the particle's _____ because of the increase in the particle's energy.

Answer: mass   Difficulty: II   Section: 4   Objective: 3

60. A span of thirty seconds is considered a _____ time for the existence of a superheavy element.

Answer: long   Difficulty: I   Section: 4   Objective: 3

SHORT ANSWER

61. What was John Newland's contribution to the development of the modern periodic table?

Answer: He was the first to arrange elements according to their chemical and physical properties and in order of increasing mass.

Difficulty: I   Section: 1   Objective: 1

62. Why was Mendeleev's table called a periodic table?

Answer: The properties of the elements occurred at repeated intervals within the table.

Difficulty: I   Section: 1   Objective: 1

63. State the periodic law.

Answer: Physical and chemical properties of the elements are functions of their atomic numbers.

Difficulty: I   Section: 1   Objective: 2

64. An element that has an electron configuration of $[Na]3s^23p^4$ is in what period and group in the periodic table?

Answer: Period 3, Group 16

Difficulty: II   Section: 2   Objective: 1

65. What is unique about the nuclei of all elements of the actinide series?

Answer: They are all radioactive; that is, they are unstable and break apart.

Difficulty: II   Section: 2   Objective: 2

66. What is the atomic number and atomic mass number of the element that is formed in the following nuclear reaction that takes place in the sun?

$$^4_2He \text{ nucleus} + {}^8_4He \text{ nucleus} \rightarrow X + \gamma \text{ radiation}$$

Answer: atomic number, 6; atomic mass number, 12

Difficulty: II   Section: 4   Objective: 1

67. In going down a group in the periodic table, what effect does electron shielding generally have on the effective nuclear charge acting on the outermost electron in an atom?

Answer: Electron shielding decreases the effective nuclear charge.

Difficulty: I   Section: 4   Objective: 2

68. In going left to right across a period in the periodic table, what effect does electron shielding generally have on the effective nuclear charge acting on the outermost electron in an atom?

    Answer: Electron shielding has little effect on the effective nuclear charge.

    Difficulty: I   Section: 4   Objective: 2

69. Explain why hydrogen is unique among all the elements.

    Answer: Hydrogen has only one proton and one electron. This condition allows it to react with many other elements. It forms water with oxygen. It reacts with nitrogen to form ammonia and with carbon to form the molecules that are essential for life.

    Difficulty: I   Section: 4   Objective: 2

70. Helium has only two electrons. Why does it behave as a noble gas?

    Answer: Helium has two electrons that completely fill the first energy shell. This makes helium very stable and nonreactive, just like the other elements in the noble gas group.

    Difficulty: I   Section: 4   Objective: 2

ESSAY QUESTIONS

71. Explain why hydrogen behaves differently from other elements, and give some examples of the kinds of compounds hydrogen forms.

    Answer:

    Hydrogen has only one proton and one electron. This condition allows it to react with many other elements. It forms water with oxygen. It reacts with nitrogen to form ammonia and with carbon to form the molecules that are essential for life.

    Difficulty: II   Section: 2   Objective: 1

72. A factory in your town uses cadmium as part of its manufacturing process. Due to recent economic conditions, cadmium is no longer available, and you must find a replacement. Suggest two elements that might be good replacements. Which of these would be your choice? Explain why.

    Answer:

    To replace cadmium, you would need an element with similar chemical properties. Zinc and mercury are in the same group as cadmium, so they would be likely choices. Mercury is a liquid, which may not be desirable, and it is also poisonous, so zinc would be the preferred replacement.

    Difficulty: II   Section: 2   Objective: 2

73. Explain how electron shielding affects the general trend in radii of atoms of elements going from left to right across a period in the periodic table.

    Answer:

    Electron shielding has little effect on the effective nuclear charge because electrons are being added to principal energy levels, not to any inner levels. As the nuclear charge increases across a period, the effective nuclear charge acting on the outermost electrons increases. The increased effective nuclear charge pulls the outermost electrons closer and closer to the nucleus and thus reduces the radius of the atom.

    Difficulty: II   Section: 3   Objective: 2

74. Compare the role of nuclear fusion in the formation of Earth's naturally occurring elements and in the creation of synthetic elements in the laboratory.

Answer:

Naturally occurring elements are created through nuclear fusion in the interior of stars. Nuclear fusion occurs in stars when the nuclei of two or more atoms join together to form the nucleus of a larger atom. Nuclear fusion of synthetic elements can be accomplished with particle accelerators, machines in which nuclei are bombarded with particles. The particles fuse with the nuclei to create nuclei of new elements.

Difficulty: II   Section: 4   Objective: 2

PROBLEMS

75. Arrange the following elements in order of increasing electron affinity: Cl, Se, S, Cs, and Te.

Answer: In increasing order: Cs, Te, Se, S, Cl

Difficulty: II   Section: 3   Objective: 4

76. Arrange the following anions (negative ions) in order of increasing ionic radius: Cl^-, F^-, Se^{2-}, and Te^{2-}

Answer: In increasing order: F^-, Cl^-, Se^{2-}, Te^{2-}

Difficulty: I   Section: 3   Objective: 4

77. Arrange the following cations (positive ions) in order of increasing ionic radius: Be^{2+}, K^+, Mg^{2+}, and Rb^+

Answer: In increasing order: Be^{2+}, Mg^{2+}, K^+, Rb^+

Difficulty: II   Section: 3   Objective: 4